University of East London
Library

Gunnar Backstrom

Practical Mathematics using MATLAB®

AF616333

Studentlitteratur

NE 9462945 5

MATLAB® is a registered trademark of The MathWorks, Inc.

British Library Cataloguing in Publication Data
A catalogue record for this book is available from the British Library

All rights reserved. No part of this publication may be reproduced or transmitted in any form or by any means, electronic or mechanical, including photocopying, recording, or any information storage and retrieval system, without permission in writing from the publisher.

Art. No. 6592
ISBN 91-44-01552-6 Studentlitteratur
© Gunnar Backstrom and Studentlitteratur 1997, 2000
Second edition

Printed in Sweden
Studentlitteratur, Lund

Printing/year 2 3 4 5 6 7 8 9 10 | 04 03 02 01

Preface

When I attended my first university courses in mathematics, the only aids available for probing the theses learnt were pen and paper, a slide rule, and a table of logarithms. Longhand calculations did not appeal to the impatient young mind, and the risk of making a trivial mistake on the way through a long transformation was disheartening. This lack of means for verifying essential results deterred many of us from fully appreciating the beautiful structure of the subject.

A mathematical theorem comes to life when you confirm it or try to falsify it. Numerical applications also shed new light on the subject, since testing a proposition requires correct understanding of what it means in terms of elementary operations.

To most students, mathematics is a means to an end, not just an admirable chain of logical deductions. The real incentive for learning the subject, at great pains, is the potential application to the facts of reality. For this purpose you need facilities for performing intricate calculations in an orderly manner and at great speed.

The advent of the pocket calculator, followed by its programmable cousin, meant the first great revolution in private-life computations. The next one occurred when affordable desk computers began to be marketed.

Today, any university student can have permanent access to a personal computer. Programs for scientific computing abound, and the main problem is to choose among the many alternatives. The selection is not easy for an individual without prior experience. In most cases the decision is dictated by academic teachers, who may be strongly influenced by their research preferences.

FORTRAN has been patched through the decades to survive into the era of the PC. This venerable programming language was originally intended for scientific computing and hence comprises exponentiation and complex numbers. The source code must be compiled and linked, however, and although much has been achieved recently to manage these procedures from the editor program, the time taken to produce executable code is excessive.

BASIC came with the first PCs and brought simplicity to programming. Here, every program line is interpreted and executed as it is entered, which greatly simplifies debugging. The logical structure has been improved and is now satisfactory, but it is difficult to add subroutines to an existing program.

PASCAL appeared on the PC scene during the 1980s. It compiles and links very quickly, and the execution is as fast as for any other language. To the casual user its obsession with data types is a serious obstacle, however, since it insists on a list of all variables to be employed.

C-language may be considered to be a development of Pascal, still insisting on type declarations of all variables and lacking rather elementary arithmetic facilities. The fact that it reaches Assembler level is no great consolation to the scientist who wishes to apply mathematics.

MATLAB® originally arose from an endeavor to simplify matrix arithmetic. It requires neither types nor dimensions, which saves much time in programming where vectors and matrices play a dominant role. It is an interpretative language, however, and hence inherently slower than FORTRAN, Pascal or C. This disadvantage is offset by the power of each statement, which may comprise a thousand elementary instructions. It also permits the code to be run line by line, which simplifies program creation and debugging.

MATLAB is an extremely practical software package, comprising both exponentiation and complex numbers. In addition, it offers convenient graphical representation of data in the form of curves and surfaces, including default scaling of the coordinate axes.

Although MATLAB is a high-level language, with many advanced procedures and functions, it is easy to overlook that it is also convenient for beginners. One only has to take care not to confuse the student by evoking the plethora of niceties that he does not need.

The idea behind the present volume is to apply the main results of undergraduate mathematics, starting with only a handful of MATLAB statements and gradually introducing more, as the need arises. When choosing the objects for calculation I have been guided by the usefulness of various operations in scientific analysis.

I wish to thank my old friend Dr. Russell Ross, formerly of the University of East Anglia, U.K., for carefully reading this volume and testing the examples.

Gunnar Backstrom

gunnar.backstrom@physics.umu.se
http://www.phys.umu.se/gunnar/

Contents

1 Introduction

In this book we are going to explore areas of mathematics that are of particular interest to scientists and engineers. Many of us do not spontaneously appreciate the abstract beauty of mathematics. We are only carried away if we discover that its results can be used for a purpose, such as predicting what is going to happen, or is likely to happen, in a given situation.

The ancient Greek mathematicians mainly concerned themselves with the objects and necessities of everyday life. Geometry served as a tool for the design and construction of buildings, for drawing maps and establishing real-estate ownership. During recent centuries, and the 1900s in particular, mathematics has developed into an extremely sophisticated structure, which no longer exhibits an obvious connection to physical reality. Even so-called applied mathematics is now dressed in such complex attire that it tends to repel those who might be interested in using it as a tool for the scientific or engineering disciplines.

The second half of the 1900s brought computers and computer programs for convenient application of mathematics, but too few mathematicians realize what a revolution it means to have such powerful tools available on one's own desk. A simple PC now permits us to illustrate the concepts of mathematics in a concrete way and to convince ourselves of the validity of theorems. We now have access to an implement for making mathematics more interesting and less difficult. Illustrating this advantage is the main purpose of the present volume.

The following examples and exercises should be regarded as a complement, rather than an alternative, to your ordinary textbooks in the various branches of mathematics. This means that you will have to consult such books for definitions and theorems from time to time. Since the majority of first-year books are much the same as regards their main contents, we need not refer to specific authors or titles.

Throughout the explorations we shall use a program package known as MATLAB®. If you have heard that this software is very advanced and complicated, forget that. You can take a ride in an airplane without understanding the significance of all the switches, instruments and signal lamps in the cockpit. As we proceed you will certainly be acquainted with more and more of the nice, special features of MATLAB, but only as the need arises. The software is of no interest in itself - unless you are selling it. We shall use it exclusively as a tool, to convince ourselves that mathematics really works. The goal is to illustrate mathematics by numeric means in such a way that the connection to the four elementary rules of arithmetic becomes evident, even in calculus.

In order to use more than the most primitive elements of MATLAB we shall need to combine sequences of commands into lists, which is usually referred to as programming. This procedure is rather similar in all programming languages and software packages, which means that the skill acquired here will become useful in later work with other products.

Programming obviously requires knowledge of the syntax, i.e. the commands and the rules for their use. This book presents the elements of syntax gradually, so that nothing needs to be known in the beginning. New commands are introduced only as required.

The examples and exercises in this volume involve some elementary numerical algorithms. These are also introduced in small portions and explained just before they are applied.

Only when the mathematical structure has been illuminated to expose its numeric meaning do we present the symbolic sister program, which exploits the core of Maple®. This part of the software will permit you to make subtle transformations of complicated mathematical expressions without paper and pencil.

MATLAB 5 runs under Windows 95/98 or Windows NT, but it is also available for the Macintosh and for workstations, the operation being similar from the user's point of view.

The examples given in this book were developed under Windows. If you are using a Macintosh or a workstation you may occasionally have to refer to your MATLAB User's Guide. The differences lie mainly in the installation procedure and in file handling.

Inexpensive Student Editions of MATLAB 5 for both Windows and Macintosh are marketed by Prentice Hall, Inc, and are obtainable in bookshops. In the present book we use the Student Edition for Windows 95/98, but the professional edition will of course work equally well.

A new Student Version (5.3) is available at campus bookstores, but only in the United States and Canada. Here, the size limits of vectors and matrices have been eliminated.

Installing the Student Edition (Version) of MATLAB

The software package is available on a CDROM and the installation is straightforward. Remember to include *help* and the *symbolic* part of the program. After the installation is finished, it is advisable to copy the MATLAB icon onto the desktop by a *shortcut.* Just double-click on that icon to start the program. This should result in a new screen with a few lines of text, followed by a "much larger than" symbol, as shown below. If that is the case, you are ready to start.

```
To get started, type one of these commands: helpwin, helpdesk, or demo
EDU»
```

(If you are using Windows without a modem for the telephone line, you might receive comments on the screen about the lack of *MIPC*. The MATLAB program will work anyway, but if you install the Windows dial-up protocol, operations become a little simpler. Advice is available on typing *help mipc*.)

Page references

As a reader I have found that textbooks generally contain much unnecessary numbering, which is not only redundant but makes the text more difficult to read. This made me choose a minimalist system for internal references. Strictly speaking, the page number is the only coordinate required. Hence, I have refrained from numbering figures and equations. Most often, there is only one figure on a page, and in any case the risk of ambiguity is exceedingly small. The bullet symbol (●) marks equations of such importance that they will be referred to later. Another simplification is that figures are discussed in the current text, just before or after the figure.

How to use this Book

This volume aspires to confer understanding of mathematics by visual and intuitive means. It also aims to instill confidence in mathematics, but by numerical verifications rather than by proofs. Detailed proofs are strictly an affair between the Mathematician and his Creator.

The research material exploited in this work dates from before 1900. The main theorems are products of giant minds, and academics have since verified them in a variety of ways. No one seriously doubts the validity of this mature part of mathematics. It could thus be taken as our cultural heritage, not to be questioned or revised.

The problem is how to incorporate mathematics into one's stock of knowledge and skill. The author believes the answer is *learning by doing*. The doing need not extend to first principles, however. To save time, one should compile blocks of knowledge, which later can serve as tools for more advanced tasks. It is important to inspect the structure of a tool, in order to gain confidence, but each tool can soon be regarded as a *black box*. The next step is to understand what the tool can do for you, and here simple, practical demonstrations are preeminent.

Here, we learn to master mathematics by means of the MATLAB program package, which contains a large number of tools, in the form of black boxes. The essential purpose is to verify that these tools produce results in agreement with known results of mathematics.

Here are a few principles to keep in mind while working with this book.

- Do not be afraid of making errors. The computer does not go up in smoke, and by correcting errors you learn the syntax of MATLAB.
- It is not enough to *read* the book. It shows a large number of results, but running the files yields additional figures that are essential to understanding.
- Typing command lines and files is a way of learning. These lists express the principles and practice of mathematical verification, and the command words constitute the language of MATLAB.

- As you work with script files, take the opportunity of displaying more than is normally shown. A statement closed by a semicolon does not display numerical values, but you can gain more insight by peeking at intermediary results. When you have learnt what is hidden, close the statement again.
- Be inquisitive! Do not accept MATLAB statements, until you understand what they do. Make a small test of your own, on one line if possible, otherwise by a short script file. Occasionally modify examples in the book to watch the effect.
- Allot time for the exercises. Some of them are simple variations of examples in the same chapter. Others expect you to be more independent. The important thing is to be active.
- Learn by doing!

2 MATLAB as a Calculator

A personal computer cannot compete with the small size of the pocket calculator, but it has certain other advantages. Anyway, we shall use it as a simple calculator in this and the following chapter, while learning the most frequent rules of the MATLAB syntax.

Many over-ambitious people make the mistake of trying to learn all the statements and rules of a programming language before applying it to specific problems. Some of them may even succeed, but in spite of their strategy. The intention with this book is to introduce the tools of MATLAB in small portions, as we illustrate the essentials of freshman mathematics. By the time you come to the last chapters, the syntax rules will be quite transparent.

Now start MATLAB by doubleclicking on the colorful triangular icon. After a few seconds the Student Edition will make itself known by the following symbol (*prompt*) on the screen

```
EDU»
```

while the professional edition simply displays the symbol », which means: "Waiting for your commands". The only rules you need for the moment are that * stands for multiplication, / for division and that 3^4 means "3 raised to the fourth power". Start at once by typing

```
3+4*5
```

and finish by pushing the <*return*> (or <↵>) key. Was the result correct? The rules are the same as for most pocket calculators. The program carries out the operations * and / before any additions or subtractions. Thus the result of

```
(3+4)*5
```

will be different, since here we force the program to calculate the contents of the parentheses separately. If you are not sure, use parentheses for clarity, even if they might not be necessary.

If there are several * and / in an expression, MATLAB begins the evaluation from the left. This implies that

```
64/8/2
```

is taken to mean (64/8)/2. Verify that by typing

```
64/(8/2)
```

and then

```
(64/8)/2
```

If an expression should also contain a number raised to a power, that operation will be carried out first of all. The result of

```
4*2^3
```

should hence be the same as

```
4*(2^3)
```

On the other hand,

```
(4*2)^3
```

is completely different.

We can always override the rules of MATLAB, by using parentheses. For instance, we may type

```
2^(3*4)
```

which also is a perfectly legal operation.

In summary, raising to a power has the highest priority, followed by multiplication and division. Addition and subtraction have the lowest priority. Within each of the three priority groups, MATLAB proceeds from left to right.

Contrary to popular belief, mathematics does not tell us anything about the real world. Mathematicians prove their theorems starting from certain axioms that are not proved. The sciences, such as physics, chemistry and biology, exploit mathematics as a *tool* for formulating theories, which in turn must be confirmed or falsified by experiment. Hence, it is the scientific disciplines that carry the ultimate responsibility for the mathematical version of their description of reality. A theory may consist of a formula or an equation that predicts what should happen in a given situation. It is important, however, that the formalisms that constitute the scientific theories are interpreted as intended. For this reason alone, it is essential to master the rules for the evaluation of arithmetic expressions.

To give a very simple example, the formula

$$r = \frac{a}{bc}$$

is unambiguous and is calculated as a/(b*c) or a/b/c. On the other hand, one has to be cautious when interpreting an expression such as

a/bc

Should we understand the sloping slash as being equivalent to the horizontal one, or does the expression mean a/b*c? If the author did not put parentheses around bc, we must assume that he meant a/b*c.

Some Useful Functions

MATLAB offers all the functions available in a pocket calculator, and many others. Powers such as a^n we have already used, for instance 2^3. In the same way, we may calculate non-integral powers as a^x by the command

```
1.23^2.31
```

The important number π is ready for use, which means that we can simply type

```
pi^2.31
```

If you should like to see this result with more decimals, you just type the command

```
format long
```

(followed by <*return*>) and hence MATLAB will print results with full precision. Note that the program always calculates using the higher precision, whether all the digits are displayed on the screen or not.

You may now obtain the preceding result with the full number of digits by (twice) pushing the arrow key ⇧ on the right hand side of the board. The line containing the power expression then reappears, and you may repeat the calculation by pushing the <*return*> key.

The square root, which is a special case of a power, is so often used that it has an alternative command, i.e. sqrt(x). You may try it by

```
sq= sqrt(2)
```

Here, we calculate $\sqrt{2}$, and at the same time we store the result in the variable sq. We may thus easily verify the square root by the inverse operation

```
sq^2
```

The function e^x is so frequently used in science that it has been given a specific notation, exp(x), which also is part of the MATLAB program. For instance, let us test

```
e=exp(1)
```

The exponential base, *e*, is not pre-defined in MATLAB, but by this assignment we have just included it. The variable name stores the value as long as you do not re-define *e* or erase the value by closing the program. This means that if you now type

```
e
```

the number will be shown again.

To convince ourselves that the number *e* is correct, we can verify it by the inverse calculation

```
log(e)
```

Curiously, the natural logarithm is named *log* in MATLAB, rather than *ln*. All such details about definitions are available by the *help* command. For instance,

```
help exp
```

not only renders information about exp(x), but also tips us about other functions of the same family, such as log10(x). We may test the latter by

```
log10(100)
```

The line

```
exp(-10)
```

yields a long number, where the final group e-005 means 10^{-5}.

The common trigonometric functions are also part of the MATLAB arsenal, and we only have to type

```
s=sin(1)
```

to obtain a very precise value for the sine of 1 radian. To check this result, we may again use the inverse function as follows

```
a=asin(s)
```

The MATLAB notation asin is a shorter version of the standard name arcsin(x) commonly used in pure mathematics. We may summarize the above test by the line

```
s1=sin(a), s, error=s-s1
```

Here, we have placed three commands on the same line, separated by commas, but we could also have used one line for each of them, as before.

Another interesting suite of commands is

```
p=2*asin(1), pi, error=p-pi
```

which demonstrates a way of generating the number π and of testing the result. We can also verify the precision by

```
cos(p), sin(p)
```

but as you see, the result in the latter case is not exactly zero. This is something you have to get used to, since numerical calculations usually have limited precision. The long format typically gives us 14 decimals, however, which is far more than normally required in practice.

The trigonometric functions in MATLAB assume that the angle is given in radians. If we prefer degrees, we must convert to radians before using the functions, e.g. by typing

```
sin(3 /180*pi)
```

which yields the sine of 3°.

The common trigonometric functions sin(x), cos(x) and tan(x) are usually defined by means of the unit circle, and of course these functions take the same value if we increase or decrease x by 2π. Furthermore, it is obvious that $\sin(x) = \sin(\pi - x)$. Let us study this point by the following line of commands.

```
x=0.4, s1=sin(x), s2=sin(pi-x), s3=sin(x+2*pi)
```

The recurring values of a trigonometric function imply that the corresponding *inverse* function, which yields the angle on the unit circle, will assume several values for a given value of the direct function. In a similar way, we may illustrate this fact by the line

```
x=0.4, s=sin(x), a1=asin(s), a2=pi-a1, a3=a1-2*pi, a4=a1+2*pi
```

We may verify that the values a_j really are equivalent by typing one more line.

```
s1=sin(a1), s2=sin(a2), s3=sin(a3), s4=sin(a4)
```

Evidently, the results are identical, down to the 14th decimal.

The lesson to be learnt from this example is that the inverse functions asin, acos and atan only provide a basic value for the angle, and we must be prepared to find the alternative angles ourselves - why not by MATLAB? Of course, it would have been impractical if the program had presented an infinite number of answers!

Exercises

❑ Compare the results of 2*3/5/7, 3*2/5/7, 2/5*3/7, and 2/7/5*3.

Evaluate the following expressions without parentheses, then with parentheses to define in what order you think the calculations should proceed. Compare the results.

❑ $4+5/2\cdot 7$ (Answer: 21.5)

❑ $3+\frac{9\cdot 4}{3\cdot 2}\cdot 8/2-5$ (22.0)

❑ $3+\frac{9\cdot 4\cdot 2^3}{3.3}\cdot 8/2+5^{\pi}$ (509.08)

❑ $3.5-2/17\cdot 3^{2\pi}$ (-113.56)

❑ $\sin(1.5)/\exp(2)\cdot 2$ (0.27)

❑ $\frac{\sin(25^{\circ})e^{-3.1}}{\exp(2.3)}\ln(1.7)/4\pi$ (7.955e-4)

❑ Invent some examples of your own where you verify calculations by doing them in different ways.

3 Geometric Applications

We shall now proceed with calculations relevant to the real world. Measurable quantities, e.g. lengths, must then be expressed in terms of a *unit*. In this book we shall adopt the international SI system, where the units are meter, kg, and so on. If we specify a length as 3.45, it should be understood that we mean 3.45 m. Angles are in radians when no unit is given, but when an angle is expressed in degrees this will be indicated by the usual symbol (^{o}).

The Cosine Theorem

As the object for our calculations we choose a triangle, or rather an *arbitrary* triangle, since we shall not have to fix the lengths from the beginning. The notations for angles and side lengths are defined by the following figure. The angle A subtends the side a, and so on.

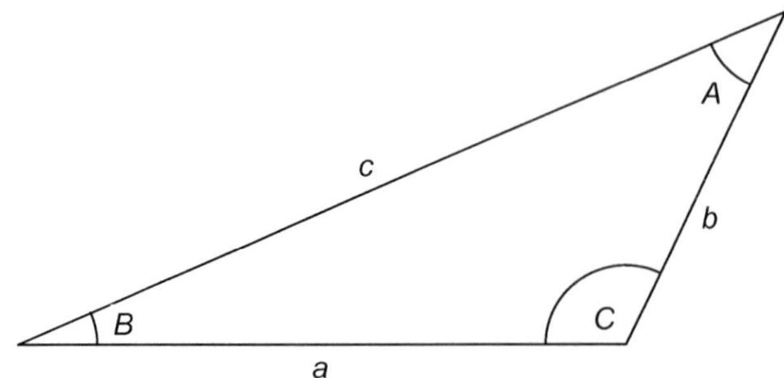

If we know two sides and the angle between them, we may calculate the missing quantities by means of the cosine theorem. If the sides a and b are known, as well as the included angle C, this theorem may be written

$$c^2 = a^2 + b^2 - 2ab\cos(C)$$ ●

In MATLAB, we first assign values to the side lengths (in meters) and to the angle (116° converted to radians), and then we calculate the side c. The program is case-sensitive, which means that the variables c and C are considered to be different.

```
a=8.5,  b=5.1,  C=116/180*pi,  c=sqrt(a^2+b^2-2*a*b*cos(C))
```

As soon as we have pushed the *return* key, all four quantities appear on the screen, and the answer becomes c= 11.6733. If we do not wish to see a known value again, we may abstain from having it displayed on the screen by using a semicolon to close the statement.

We now know the sides (a,b,c), but only one angle. The cosine theorem just yielded the side c, using the known value for the angle C. Obviously there must be similar expressions for the other angles, i.e.

$$a^2 = b^2 + c^2 - 2bc\cos(A), \quad b^2 = a^2 + c^2 - 2ac\cos(B) \qquad \bullet$$

From this we obtain the cosines of the remaining angles, and we only have to use the command acos to calculate A and B.

```
A=acos( (a^2-b^2-c^2)/(-2*b*c) ),  B=acos( (b^2-a^2-c^2)/(-2*a*c) )
```

In this case, the acos command yields unique answers, considering that the alternative values for the angles are of the opposite sign or larger than π.

Let us next express the angles in degrees (Adg, and so on) and also verify that the sum of the angles really is 180°.

```
Adg= A/pi*180,  Bdg=B/pi*180,  Cdg=C/pi*180,  sum=Adg+Bdg+Cdg
```

It is easy to repeat the above calculations with other values for a, b, and C. If we push the arrow ⇧ on the right hand side of the keyboard, we retrieve the preceding MATLAB line. Repeating this twice more we recover the line where we may enter new input values.

```
a=1.5,  b=2.4,  C=36/180*pi,  c=sqrt(a^2+b^2-2*a*b*cos(C))
```

After we have finished by *return*, we obtain the side c corresponding to the new values. In a similar way, you may calculate the other angles of this triangle and convert these to degrees. Do this on your own!

A triangle is completely defined if all sides (a,b,c) are known. By the first cosine theorem (p.12) we may calculate the angle C, and the other cosine theorems give us the remaining angles.

The Sine Theorem

Now assume that we know the sides a and b, as well as the angle A. What is the simplest way to calculate the other sides and angles? The first cosine theorem (p.12) would involve *two* unknown quantities, c and C. In this situation, the sine theorem is convenient. It reads

$$\frac{\sin(A)}{a} = \frac{\sin(B)}{b} = \frac{\sin(C)}{c}$$ ●

In our case, $\sin(A)/a$ has a known value, and the first equality then gives us $\sin(B) = b\sin(A)/a$. From that we calculate the angle B using the function arcsin(x). Hence, the angle C also becomes known from the relation $C = \pi - A - B$.

```
a=4.1, b=9.1, A=20/180*pi, sB=b*sin(A)/a, B=asin(sB), C=pi-A-B
```

The last of these commands guarantees that the sum of the angles will be π. We finally obtain the side c from the sine theorem $c = a\sin(C)/\sin(A)$. At the same time, we may express all angles in degrees.

```
c=a*sin(C)/sin(A), Adg=A/pi*180, Bdg=B/pi*180, Cdg=C/pi*180
```

At this point we must recall that the function v=arcsin(x) is multi-valued. Within the unit circle there is another, equivalent value for the angle, namely $\pi - v$. Let us explore if this alternative gives us a new triangle.

```
a=4.1, b=9.1, A=20/180*pi, sB=b*sin(A)/a, B=pi-asin(sB), C=pi-A-B
```

To find c and to convert the angles to degrees, we only need to recall the preceding line by means of the arrow key ⇧, and finish by *return*.

The result of this calculation is a distinctly different triangle. The next figure illustrates how the two cases arise. We knew the angle A and the sides (a,b), but the angle C was initially undetermined. As is evident from the figure, the side a may connect to the side c in two different ways (a and a'), which the numeric solution confirms.

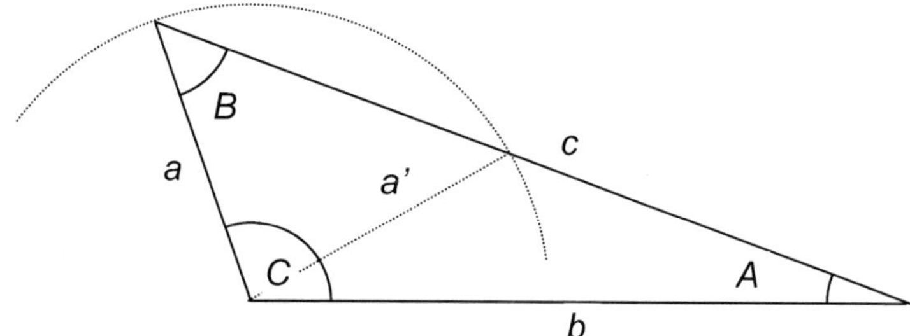

Area of a Triangle

The best-known formulae for calculating the area A of the above triangle are

$$A = \frac{1}{2}bh = \frac{1}{2}ba\sin(C)$$ ●

where b is the base, or the horizontal side, and h the height of the vertex above the base. There is also a beautiful expression that only involves the sides, which may be written

$$A = \sqrt{s(s-a)(s-b)(s-c)}$$ ●

where s is one-half of the perimeter, or $s = (a+b+c)/2$.

If we have not switched off the computer, closed MATLAB or otherwise erased the previous results, the sides and angles are still available in memory. Otherwise we must repeat the last line of operations and then recall and execute the command line for calculating c.

We may next compare the two equivalent expressions for the area (A1 and A2) by means of a single MATLAB line.

```
a, b, c, A1=0.5*b*a*sin(C), s=(a+b+c)/2; A2=sqrt(s*(s-a)*(s-b)*(s-c))
```

If we obtained the results in format short, we now enter format long and then repeat the preceding line. We see that the two values for the area actually are different in the 14th decimal. This is an example of the round-off errors that may occur in numerical calculations.

These are only a few examples of the many calculations possible with geometric objects. Whenever there is an expression for a quantity, MATLAB is at hand for easy evaluation.

Exercises

❑ A triangle has the sides $a = 0.7$, $b = 0.9$, and $c = 1.1$. Calculate the corresponding angles in degrees.

❑ A triangle has $a = 1.5$, $B = \pi/4$, and $C = 85^\circ$. Solve for the other sides and the angle A.

❑ A parallelogram has sides 0.7 and 1.1 and one of the angles is 35°. Calculate the two diagonals and the area.

❑ A regular polygon of n sides is inscribed in a circle of radius $r = 1.33$. Considering this polygon to consist of identical triangles, calculate the area and perimeter of the polygon for $n = 6$, $n = 60$, and $n = 600$. Also calculate the area and perimeter of the circle for comparison.

❑ A regular polygon with n sides has the perimeter $p = 10.6$. Calculate its area for $n = 6$, $n = 60$, and $n = 600$. Also find the ratio of the area to the perimeter and make your own comparison.

4 Vectors in (*x*, *y*) and (*x*, *y*, *z*)

An arrow from O to P on a map may represent a displacement. If we lay down a coordinate system on the map, e.g. with the x-axis pointing to the east and the y-axis to the north, we may also describe the displacement **A** as the result of two parts, A_x and A_y, measured along these two axes (next figure). The displacement **A** may hence be considered to be a *vector* in the (x, y) plane, i.e. a quantity comprising the two components A_x and A_y.

The vector **A** may also be taken as the *position* of a point in the plane, once the origin O has been defined.

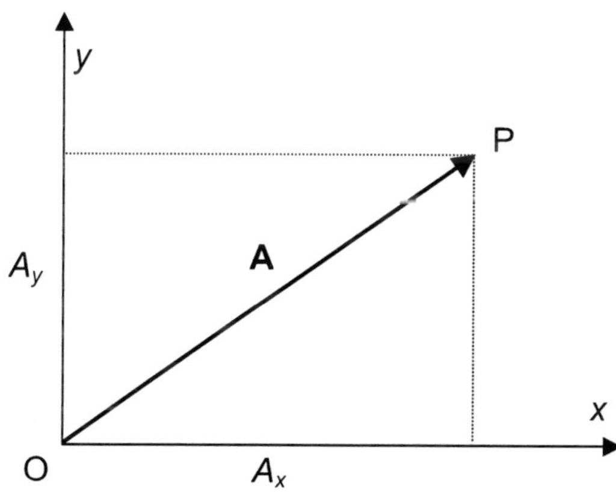

We now leave the world of simple pocket calculators. MATLAB provides a very convenient notation for vectors, and it also permits us to manipulate them in various ways. For instance, we may type

```
A=[6 2]
```

in order to store the two values A_x and A_y in the vector **A**. Evidently, the program responds to *return* by displaying these components in a row.

The magnitude $|\mathbf{A}|$ (length) of a vector **A**, is defined by

$$a \equiv |\mathbf{A}| = \sqrt{A_x^2 + A_y^2}$$

and we may compute it by the command

```
A=[6 2], a=sqrt(6^2+ 2^2)
```

We can safely use the notation a, since MATLAB is sensitive to the difference between upper- and lowercase characters. In order to distinguish vectors from scalar numbers we generally use bold uppercase for vectors.

If we now define another, similar object

```
B=[3 6], b=sqrt(3^2+ 6^2)
```

we may ask MATLAB to add or subtract these vectors. The total result of the displacement **A**, followed by the displacement **B**, may be calculated by the simple line

```
C=A+B
```

The figure below indicates how you can use a graphical method (by pencil and paper) to add **A** to **B**, or **B** to **A**, to obtain **C**.

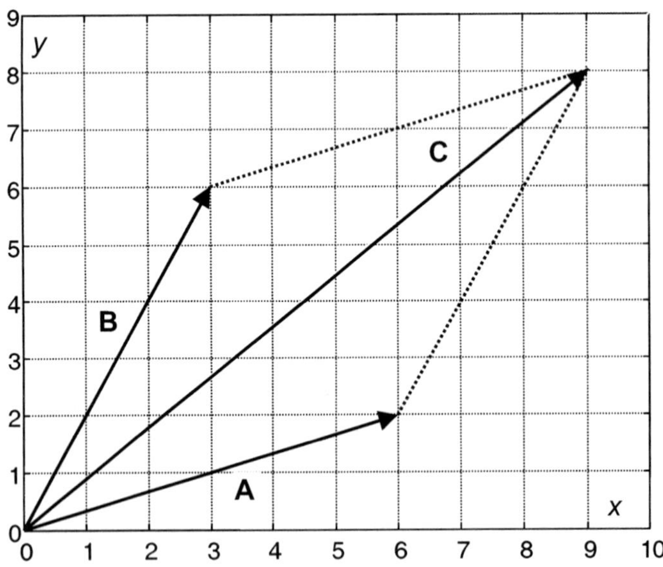

Not only can you add and subtract two vectors; you can also multiply them in different ways. The *scalar* product is defined as

$$s = \mathbf{A} \cdot \mathbf{B} \equiv ab\cos(\theta), \qquad \bullet$$

where a and b are the magnitudes of the vectors and θ the angle between them. There is a built-in function for this operation, i.e.

```
s=dot(A,B)
```

where *dot* refers to the conventional sign used for scalar multiplication of vectors.

There is an equivalent definition of the scalar product. If the angles from the x-axis to the vectors **A** and **B** are denoted θ_A and θ_B, we may rewrite the expression for the scalar product as follows.

$$s = ab\cos(\theta) = ab\cos(\theta_B - \theta_A) =$$
$$ab\{\cos(\theta_B)\cos(\theta_A) + \sin(\theta_B)\sin(\theta_A)\} =$$
$$a\cos(\theta_A)b\cos(\theta_B) + a\sin(\theta_A)b\sin(\theta_B)$$

This leads to the alternative expression

$$s = A_x B_x + A_y B_y \qquad \bullet$$

One or the other of these two definitions may be more convenient, depending on the situation.

It is worth noticing that the scalar product of a vector with itself becomes

$$\mathbf{A} \cdot \mathbf{A} \equiv aa\cos(\theta) = a^2$$

which means that we can obtain the magnitude of vectors

```
a=sqrt(dot(A,A)),  b=sqrt(dot(B,B)),  c=sqrt(dot(C,C))
```

without using the component values explicitly.

If we now compute $\mathbf{A} \cdot \mathbf{C} / (ac)$ to obtain the $\cos(\theta)$,

```
cosine=dot(A,C)/(a*c),  thetaB=acos(cosine),  thetaBdg=thetaB/pi*180
```

and then apply *acos*, we finally get $\theta = 23.2^\circ$ for the included angle.

By the same method, we may calculate other angles of the above triangle, but we must realize that the angle between two straight lines is not uniquely defined. There are in fact *two* angles at the intersection, the sum of them being 180°. In the preceding chapter we referred to the *internal* angles of a triangle, in particular when stating

94629455

that their sum is 180°. Vectors have directions, however, and in order to obtain unambiguous angles we only have to use vectors pointing *away* from the vertex.

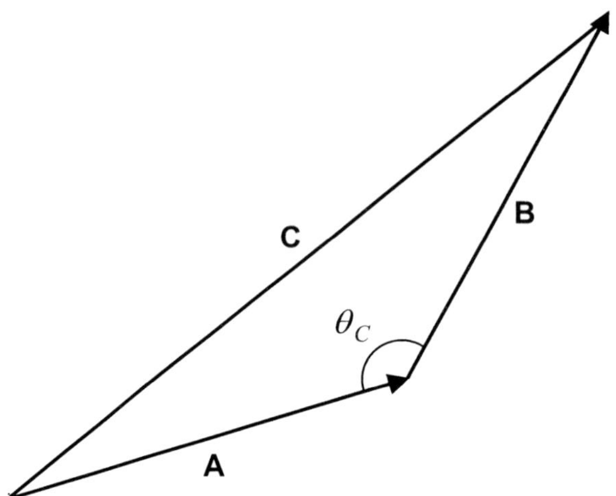

As an example, let us calculate the angles of the above triangle defined by the old vectors **A**, **B**, and **C**. We already have the side lengths, and now we pursue the calculation for each of the angles. The line

```
cosine=dot(-A,B)/(a*b), thetaC=acos(cosine), thetaCdg=thetaC/pi*180
```

gives us $\theta_C = 135^\circ$. Notice the minus sign for **A**, which is required because this vector points *into* the vertex at θ_C.

We then repeat this line of commands for the remaining angles:

```
cosine=dot(A,C)/(a*c), thetaB=acos(cosine), thetaBdg=thetaB/pi*180
```

and

```
cosine=dot(-B,-C)/(b*c), thetaA=acos(cosine), thetaAdg=thetaA/pi*180
```

Evidently, the angles agree rather well with what can be measured in the figure, and the sum of them is indeed 180°.

Vectors in (x, y, z)

Velocity is an example of a vector quantity. An arrow flying through the air moves simultaneously in the east, north and vertical directions, which means that its velocity may be given by three components. For example, the velocity components might be 5 m/s north, -3 m/s east and 1 m/s vertical, or in short [5 -3 1].

Vectors in the three dimensions of space may be defined in the same simple way by their x, y, and z components, for instance

```
A=[1 2 3], a=sqrt( dot(A,A)), B=[2 3 -1], b=sqrt( dot(B,B))
```

These vectors can be added and subtracted in the same way as vectors in two dimensions. Verify that by typing

```
C=A+B
```

We may also calculate the scalar product by exactly the same command as before, i.e.

```
s=dot(A,B)
```

This means that we again obtain the angle between **A** and **B** by the following line.

```
cosine=dot(A,B)/(a*b), theta=acos(cosine), thetadg=theta/pi*180
```

In three-dimensional space, it is possible to define a *vectorial* product (or *cross* product) **V** of the vectors **A** and **B**.

$$\mathbf{V} = \mathbf{A} \times \mathbf{B}$$

This new product, which is frequently used in both geometry and physics, is defined by

$$v \equiv |\mathbf{V}| = a\,b\sin(\theta) \qquad \bullet$$

where v is the magnitude of **V**. The latter vector is directed at right angles to the plane containing **A** and **B**, as shown by **N** in the figure below.

If both **A** and **B** are in the (x,y) plane, we may express the *cross* product by the vector components, as we did for the *dot* product (p.19).

$$\begin{aligned} v \equiv |\mathbf{V}| &= a\,b\sin(\theta) = ab\sin(\theta_B - \theta_A) = \\ &ab\{\sin(\theta_B)\cos(\theta_A) - \cos(\theta_B)\sin(\theta_A)\} = \\ &a\cos(\theta_A)b\sin(\theta_B) - a\sin(\theta_A)b\cos(\theta_B) \end{aligned}$$

$$v = A_x B_y - A_y B_x \qquad \bullet$$

If **A** and **B** are the sides of a triangle, then $v = a\,b\sin(\theta) = a\,h$, where h is the height above a. Hence, $|\mathbf{A}\times\mathbf{B}|/2 = a\,b\sin(\theta)/2$ is equal to the area of the triangle.

For any vectors in (x, y, z) space, we conveniently obtain this *cross* product by the MATLAB command

```
V=cross(A,B)
```

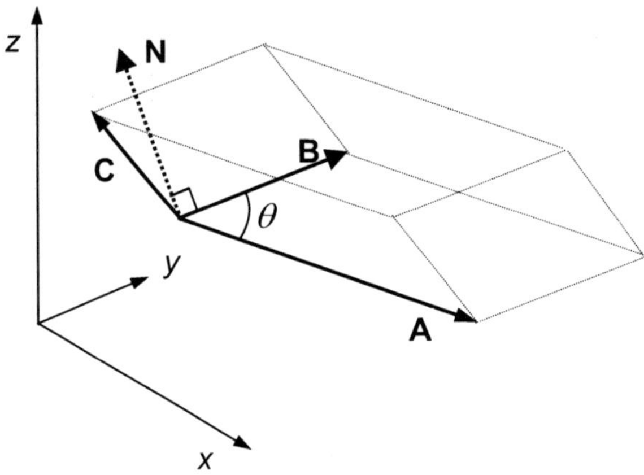

Let us consider the above parallelepiped, defined by **A**, **B**, and **C**. According to elementary geometry its volume is given by the product of the base area $|\mathbf{A}\times\mathbf{B}| = a\,b\sin(\theta)$ and the height. The latter is the projection of **C** on the vector **N,** normal to the base plane, and hence we have

volume=$\left|\mathbf{C}\cdot(\mathbf{A}\times\mathbf{B})\right|$

where we take the absolute value of the expression, to be assured of a positive value for the volume.

Thus the following three vectors **A**, **B,** and **C** generate a parallelepiped with a volume of 9 units, which you may verify by typing

```
A=[1 2 3];  B=[2 -2 -3];  C=[3 3 3];  vol=abs( dot( C, cross(A,B)))
```

The *area* of the parallelepiped is simply calculated by summing the cross products for the six faces.

Exercises

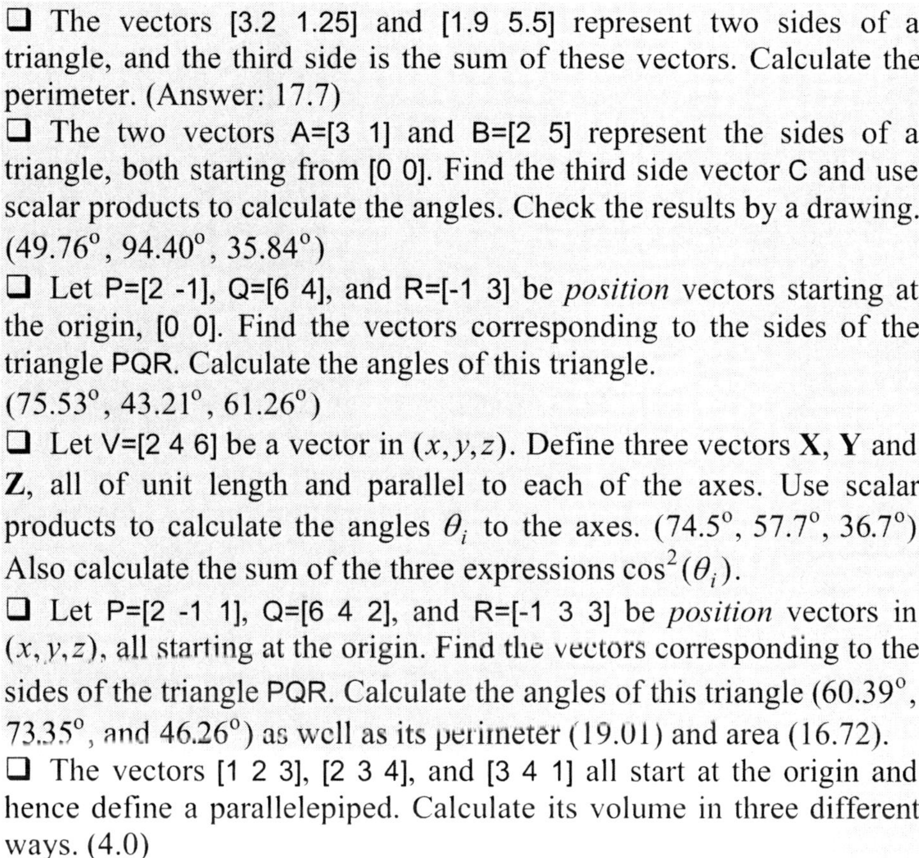

❑ The vectors [3.2 1.25] and [1.9 5.5] represent two sides of a triangle, and the third side is the sum of these vectors. Calculate the perimeter. (Answer: 17.7)

❑ The two vectors A=[3 1] and B=[2 5] represent the sides of a triangle, both starting from [0 0]. Find the third side vector C and use scalar products to calculate the angles. Check the results by a drawing. (49.76°, 94.40°, 35.84°)

❑ Let P=[2 -1], Q=[6 4], and R=[-1 3] be *position* vectors starting at the origin, [0 0]. Find the vectors corresponding to the sides of the triangle PQR. Calculate the angles of this triangle. (75.53°, 43.21°, 61.26°)

❑ Let V=[2 4 6] be a vector in (x, y, z). Define three vectors **X**, **Y** and **Z**, all of unit length and parallel to each of the axes. Use scalar products to calculate the angles θ_i to the axes. (74.5°, 57.7°, 36.7°) Also calculate the sum of the three expressions $\cos^2(\theta_i)$.

❑ Let P=[2 -1 1], Q=[6 4 2], and R=[-1 3 3] be *position* vectors in (x, y, z), all starting at the origin. Find the vectors corresponding to the sides of the triangle PQR. Calculate the angles of this triangle (60.39°, 73.35°, and 46.26°) as well as its perimeter (19.01) and area (16.72).

❑ The vectors [1 2 3], [2 3 4], and [3 4 1] all start at the origin and hence define a parallelepiped. Calculate its volume in three different ways. (4.0)

5 General Vectors

Ordinary physical space has three dimensions, but a general vector may contain any number of components, or elements, as they are usually called. Since we are now going to experiment with a new set of vectors, we had better erase all previous variables by *clear*.
If we type

```
clear,  A=[1 2 3 4],  B=[5 6 7 8]
```

we have defined two vectors, each containing four integral values. We should not try to visualize such vectors in four-dimensional space, but just consider them to be objects containing numbers in a given order. In the following chapters we shall see how useful these vectors are in practical calculations.

If two vectors have the same number of elements, we may add them like this

```
A+B
```

and from the answer on the screen you will see that the new vector has the same number of elements, each of them being the sum of the original ones. Subtraction is also a valid operation and works in a corresponding fashion

```
A-B
```

as you easily conclude from the result.

Although this might be unexpected, MATLAB also permits us to add a *scalar* to a vector, or subtract it, for instance

```
A+0.5,  A-0.5
```

As you see from the results, the program assumes that we really mean to add a vector with the same number of elements as in **A**. The above line is thus a short way of typing

```
A+[0.5 0.5 0.5 0.5],  A-[0.5 0.5 0.5 0.5]
```

Furthermore, we may multiply or divide a vector by a single number (scalar), e.g.

```
C=3.33*A
```

Here, a common factor multiplies every element of the vector.

More Element-by-Element Operations

We now come to a more general multiply operation, so convenient and so frequently used in applied mathematics that MATLAB created a special symbol for it. We may call it *"point" multiplication*, as suggested by the following example

```
A, B, D=A.*B
```

This operation is really analogous to the addition of two vectors. It means that each element in the first vector is multiplied by the corresponding element in the second one. The result is a new vector, containing the same number of elements. Note that this new type of operation is different from "*dot*" multiplication (p.19), which yields a scalar result.

The "point" is small but essential in the above multiplication, since the operation A*B is not defined. To try this, recall the preceding line by the ⇧ key, remove the point and try again.

If we could define element-by-element multiplication, why not division in the same manner? Indeed, that can also be done, as shown by the following test

```
D./B
```

where the point plays an equally important role. This operation is evidently the inverse of point multiplication, since the result is equal to A.

While we are at it, why not extend the principle of element-by-element operation to powers? Again, this application is also valid in MATLAB, which you may verify by typing

```
A.^2
```

The result of this operation is evidently $[a_1^2\ a_2^2\ a_3^2\ a_4^2\ ]$, which means that each vector element a_j is squared, the squares forming a new vector.

We may proceed to arbitrary powers of A by the line

```
A.^B
```

where A contains all the bases and B the exponents. As is clear from this test, the result may be written as

$$\left[a_1^{b_1}\ a_2^{b_2}\ a_3^{b_3}\ a_4^{b_4}\right]$$

Evidently, the exponentiation is applied to corresponding elements of the first vector, and the results are placed in a new vector.

Note that the point in .^ modifies the exponentiation operator ^, not the preceding vector. Similarly, the pairs .* and ./ should be seen as modified multiply and divide operations.

We are now ready to take the final step towards a general element-by-element operation. Let us consider a *function of a vector*, e.g.

```
cos(B),  [cos(5) cos(6) cos(7) cos(8)]
```

and compare to a vector containing functions of each element. This test indicates that the result of the operation cos(B) is

$$[\cos(b_1)\ \cos(b_2)\ \cos(b_3)\ \cos(b_4)]$$

What is the practical use of a function of a vector? If you want to plot a function $f(x)$, you will start by deciding at what intervals of x you want to calculate the function. This means that you design a vector X, containing the elements $[x_1\ x_2\ x_3 \ldots x_n]$ and then calculate the function values $f(x_j)$ for each of these elements. The MATLAB element-by-element operation does this very efficiently. We shall often exploit this powerful feature in the following chapters.

Special Operations on Vectors

There are operations that are meaningless when applied to scalars but which are useful in the case of vectors. To separate a vector into its elements is such a special operation. For instance, the line

```
A(2),  A(4)
```

yields elements number 2 and 4 of the vector A. Using this notation we may for instance compute the magnitude of a 4-element vector by the line

```
a=sqrt( A(1)^2+ A(2)^2+ A(3)^2+ A(4)^2)
```

The terse command *sum* serves to add all elements of a vector. Let us first apply it to the vector A as follows.

```
A, sumA=sum(A)
```

Since this evidently works, we may also use it to compute the magnitude of a vector, i.e.

```
a=sqrt( sum( A.^2))
```

which remains valid for any number of elements in the vector.

The scalar product (p.19) in the (x, y) plane may be written

$$s = \mathbf{A} \cdot \mathbf{B} \equiv ab\cos(\theta) = A_x B_x + A_y B_y$$

For general vectors with n elements we may define $\mathbf{A} \cdot \mathbf{B}$ as

$$s_n = a_1 b_1 + a_2 b_2 + a_3 b_3 + \ldots + a_n b_n \qquad \bullet$$

We may calculate this series in two steps, first by the element-by-element multiplication

```
A, B, A.*B
```

which yields the products as a vector. To obtain the scalar product we just need to sum over these terms by

```
sn=sum(A.*B), snx=dot(A,B)
```

Evidently, the familiar *dot* command is still valid. As we shall see, the scalar product has numerous practical applications, and we may use any of these forms for it.

There are a few other operations on vectors that are occasionally useful. We may find the (algebraically) smallest and largest elements of a vector by

```
min(B), max(B)
```

and the number of elements may be accessed by

```
n=length(B)
```

Notice that length has nothing to do with the *magnitude* of a vector.

The above functions form a powerful arsenal that we shall use time and again. We have systematically employed uppercase characters for vectors. This is by no means necessary, but it helps us to be aware of what objects are vectors and to remember the rules concerning multiplication and division.

Row and Column Vectors

It is customary in the sciences to present vector elements in a row. In mathematics, however, there are two types of vectors, i.e. row and column vectors. MATLAB acknowledges the difference and displays them accordingly, as we may see by typing

```
A, Acolumn=A'
```

Here, the quote (') transforms a row to a column. In mathematics, this operation is called *transposing* a vector, and is usually denoted A^T. By transposing once more we regain the original vector, as shown by

```
A', (A')'
```

We have already noticed (p.25) that the expression A*B is not defined. On the other hand, the product A*B' is valid as we see by testing

```
A*B', snx=dot(A,B)
```

As is evident from the above comparison, the new expression is equivalent to the scalar product. Thus A*B' is the *sum* of the terms generated by an element-by-element product.

At the moment, it may be obscure why we need column vectors in the first place, but in the next chapter we shall see that they play an essential role in practical calculations.

Summary of Element-by-Element Operations

It might be useful at this point to summarize some operations peculiar to MATLAB vectors. Column vectors are particularly convenient for illustration, although similar rules apply to row vectors. Longer vectors are of course also treated similarly.

$$\mathbf{A} \pm \mathbf{B} = \begin{bmatrix} a_1 \\ a_2 \end{bmatrix} \pm \begin{bmatrix} b_1 \\ b_2 \end{bmatrix} = \begin{bmatrix} a_1 \pm b_1 \\ a_2 \pm b_2 \end{bmatrix} \qquad \mathbf{A} \pm c = \begin{bmatrix} a_1 \\ a_2 \end{bmatrix} \pm c = \begin{bmatrix} a_1 \pm c \\ a_2 \pm c \end{bmatrix}$$

$$\mathbf{A.*B} = \begin{bmatrix} a_1 \\ a_2 \end{bmatrix} .* \begin{bmatrix} b_1 \\ b_2 \end{bmatrix} = \begin{bmatrix} a_1 * b_1 \\ a_2 * b_2 \end{bmatrix} \qquad \mathbf{A} * c = \begin{bmatrix} a_1 \\ a_2 \end{bmatrix} * c = \begin{bmatrix} a_1 * c \\ a_2 * c \end{bmatrix}$$

$$\mathbf{A./B} = \begin{bmatrix} a_1 \\ a_2 \end{bmatrix} ./ \begin{bmatrix} b_1 \\ b_2 \end{bmatrix} = \begin{bmatrix} a_1/b_1 \\ a_2/b_2 \end{bmatrix} \qquad \mathbf{A}/c = \begin{bmatrix} a_1 \\ a_2 \end{bmatrix} / c = \begin{bmatrix} a_1/c \\ a_2/c \end{bmatrix}$$

$$\mathbf{A.\wedge B} = \begin{bmatrix} a_1 \\ a_2 \end{bmatrix} .\wedge \begin{bmatrix} b_1 \\ b_2 \end{bmatrix} = \begin{bmatrix} a_1 \wedge b_1 \\ a_2 \wedge b_2 \end{bmatrix}$$

Exercises

Erase previous definitions by *clear* and create the three vectors
A=[1 2 3 4 5], B=[4 5 6 7 8], C=[1 -1 2 -1 1]
for use in the following exercises.

- ❑ Calculate the vector 3A+B.
- ❑ Find the magnitude of A by three different expressions.
- ❑ Multiply A by B to obtain the vector D, and then divide D by A.
- ❑ Calculate the scalar product of A and B by three different operations.
- ❑ Try the operation A'*B'.
- ❑ Calculate sin(A)./A.
- ❑ Find the smallest and the largest element of the vector A.*C.
- ❑ Evidently, the MATLAB expression sin(A(length(A))) is valid, but what does it mean?
- ❑ Calculate (A.*B).*C and compare to A.*(B.*C). Also test if the expression A.*B.*C is allowed.
- ❑ Calculate sin(A)./cos(A)./tan(A) and verify the result.

6 Systems of Linear Equations

Empirical data usually must be brought into digestible form before they are presented graphically, and eventually interpreted. Some of the steps taken lead to systems of linear equations, which are as easy to grasp as they are tedious to solve. In this chapter we shall learn how to solve systems of a hundred unknowns without tears.

Two Unknown Variables

The procedure used for solving giant systems is the same as with two unknowns, so why not begin with the simplest possible case? The system of linear equations

$$\begin{cases} x + 2y = 8 \\ 3x + 4y = 18 \end{cases}$$

reminds us of objects we discussed in the preceding chapter. The left sides have the form of scalar products, and the right sides together form a column vector. The above system of equations may hence be written as an equality between two column vectors, which implies that corresponding elements must be equal.

$$\begin{bmatrix} 1 \cdot x + 2 \cdot y \\ 3 \cdot x + 4 \cdot y \end{bmatrix} = \begin{bmatrix} 8 \\ 18 \end{bmatrix}$$

We may rewrite the elements on the left side as scalar products (after the model of A*B' or dot(A,B) on p.28)

$$[1 \quad 2]\begin{bmatrix} x \\ y \end{bmatrix} = x + 2y, \qquad [3 \quad 4]\begin{bmatrix} x \\ y \end{bmatrix} = 3x + 4y$$

Hence, we may combine these two expressions into the following symbolic form for the above system of linear equations.

$$\begin{bmatrix} 1 & 2 \\ 3 & 4 \end{bmatrix} \begin{bmatrix} x \\ y \end{bmatrix} = \begin{bmatrix} 8 \\ 18 \end{bmatrix}$$

The new, square array is known as a *matrix*, which is to be handled according to certain rules. In fact, a row or a column vector may be seen as a special case of a matrix.

The matrix may be typed onto the MATLAB window much as we did with vectors, i.e.

```
clear,  A=[1 2; 3 4]
```

where we first *clear* earlier assignments. The semicolon inside the square brackets separates the two rows.

We already know how to type a row vector and we easily transform that into a column vector as follows.

```
B=[8 18]'
```

The crucial sign here is the *quote* after the row vector, which transposes it into a column vector. If we combine the unknown variables into a column vector X, we could also write the above system in the short form

AX=B

We may now find the solution simply by typing the terse command

```
X= A\B
```
●

which suggests that we somehow *divide* B by A to obtain X.

This example is of course trivial, since we could equally well have solved it by hand. The solution proceeds as easily, however, with a larger number of equations.

Three Unknown Variables

As a transition to the case of a large number of unknowns, let us now solve a system with three unknowns.

$$\begin{cases} 2.2x + 5.5y + 4.4z = 12.2 \\ 5.5x + 6.6y + 7.7z = 13.3 \\ 5.5x + 7.7y + 9.9z = 14.4 \end{cases}$$

The corresponding matrix form will be

$$\begin{bmatrix} 2.2 & 5.5 & 4.4 \\ 5.5 & 6.6 & 7.7 \\ 5.5 & 7.7 & 9.9 \end{bmatrix} \begin{bmatrix} x \\ y \\ z \end{bmatrix} = \begin{bmatrix} 12.2 \\ 13.3 \\ 14.4 \end{bmatrix}$$

The MATLAB input can be written

```
clear,  A=[2.2 5.5 4.4; 5.5 6.6 7.7; 5.5 7.7 9.9],  B=[12.2 13.3 14.4]'
```

We obtain a high-precision solution by the familiar command

```
format long,  X= A\B
```

With two unknowns, we could easily verify the result by pencil and paper, but for three equations that would be tedious. It is better to use MATLAB to compare AX with B. The following line is what we need.

```
left=A*X,  right=B,  error=left-right
```

As we see, the two sides balance down to the last digit displayed, but taking the difference we find that there is an error of a fraction of a unit in the last decimal. In general, we could of course not expect exact solutions from numerical calculations.

In very rare cases, we may receive a warning about large errors in the results. This happens when the values of the matrix elements are unfavorable, and in extreme cases it may even be impossible to obtain a solution. In any case, MATLAB uses the most sophisticated methods available for solving a system of equations.

Many Unknown Variables

We may experiment with large systems of equations without typing an enormous number of matrix and vector elements. MATLAB offers the command *rand*, which creates a matrix of random elements, as follows.

```
format short, clear, A=rand(7,7)
```

As you see, the program creates a 7×7 matrix containing random number elements in the range $0 \ldots 1$.

Similarly, we may create a column vector of random elements by

```
B=rand(7,1)
```

By the last statement we have created a vector with 7 rows and 1 column. We may solve this system of equations as before by typing

```
X=A\B
```

The remaining problem is again to convince ourselves that the solution is as good as can be obtained. For this purpose we employ the previous test, i.e.

```
left=A*X, right=B, error=left-right
```

The difference between the left and right side elements turns out to be in the region of 10^{-15}, which certainly is an impressive accuracy.

Matrix Algebra (Optional)

In this book we shall not need more advanced matrix algebra than the multiplication AX used in the preceding sections. Matrices have found many uses, however, in various branches of engineering and science and some readers might be interested in a preview of the surprisingly easy way that MATLAB handles these objects.

We have just multiplied a matrix and a column vector, which is a special case of general matrix multiplication. The matrices below illustrate the multiplication AB=C.

$$\begin{bmatrix} 1 & 2 & 3 \\ 2 & 1 & 1 \\ 1 & 0 & 1 \end{bmatrix} \begin{bmatrix} 3 & 2 & 3 \\ 1 & 2 & 2 \\ 3 & 1 & 2 \end{bmatrix} = \begin{bmatrix} 14 & 9 & 13 \\ 10 & 7 & 10 \\ 6 & 3 & 5 \end{bmatrix}$$

Each of these matrices may be regarded as an array of three row vectors, but equally well as an array of three column vectors. We may identify an element by its row and column numbers, e.g. b_{11} for the element in the upper-left corner of B.

We obtain the elements in the product matrix by the same procedure as when we multiplied row and column vectors. For instance, dot-multiplying the 2nd row (shaded) of the first matrix with the 1st column (shaded) of the second matrix we obtain a scalar result as follows.

$$[2 \quad 1 \quad 1]\begin{bmatrix}3\\1\\3\end{bmatrix} = 2\cdot 3 + 1\cdot 1 + 1\cdot 3 = 10$$

There are nine possible products of this kind, and we place those in the positions given by the row and column numbers, here in c_{21}.

Scalar products of this kind require that the row vector be as long as the column vector is high. On p.25 we tried to multiply two row vectors by A*B but received the error message

```
??? Error using ==> *
Inner matrix dimensions must agree.
```

The first object certainly was a row vector, and if we wish, we could view the second row vector as a matrix of size 1×3. The height of each column was only *one* element instead of three, however, and that is why the multiplication A*B did not succeed. We may always check the number of elements in the vertical and horizontal directions by the command **size**(A).

On the other hand, the operations A.*B and dot(A,B) both worked. Point-multiplication only requires that the numbers of elements in both dimensions are equal. Even matrices yield a point product, but only if the sizes in both dimensions are equal.

With two different square matrices, defined by

```
clear, A=rand(7,7); B=rand(7,7);
```

we may form two products

```
AB=A*B, BA=B*A
```

and (surprisingly) we find the resulting matrices to be different. For two scalar numbers *a* and *b*, we always find a*b=b*a, but for matrices this is not true in general.

There is an important exception to this rule. Whenever A is such that the system AX=B has a solution, we may find an *inverse* matrix A^{-1}, such that

$$A\,A^{-1} = A^{-1}A = I$$ ●

where I is known as the *identity matrix*. An identity matrix with 3×3 elements may be written

$$\begin{bmatrix} 1 & 0 & 0 \\ 0 & 1 & 0 \\ 0 & 0 & 1 \end{bmatrix}$$

Evidently, this matrix has zeros everywhere except along the descending diagonal, where the elements are unity.

MATLAB delivers the inverse matrix on the simple command

```
Ainv=inv(A)
```

It is easy to test if the inversion has been carried out correctly. Just type

```
format short,  Ainv*A,  A*Ainv
```

and watch the results. Using the short format you may get the impression that each result is an exact identity matrix, but in numerical calculations there are inevitably some round-off errors. We would see more details after changing to format long, but it is easier to compute the difference between these two "identity" matrices:

```
Ainv*A- A*Ainv
```

The difference indicates that the diagonal elements are not exactly 1, nor are the others exactly zero. The errors are only parts of 10^{-14}, however, which is acceptable for most practical calculations.

The inverse matrix is a useful object. If we multiply both sides of the equation

AX=B

from the left by the inverse matrix A^{-1} it becomes

$A^{-1}AX= A^{-1}B$

or

$IX= A^{-1}B$

but since IX=X we finally obtain

$X= A^{-1}B$

This is an alternative way of solving a system of linear equations.

The MATLAB formula, X= A\B, involves the "backslash" symbol. Even if this is just a shorthand notation, which need not really be explained, we can see that it is very well chosen. The backslash may be regarded as a division symbol, indicating that we divide B by A from the left, hence multiply by the inverse from the left.

Exercises

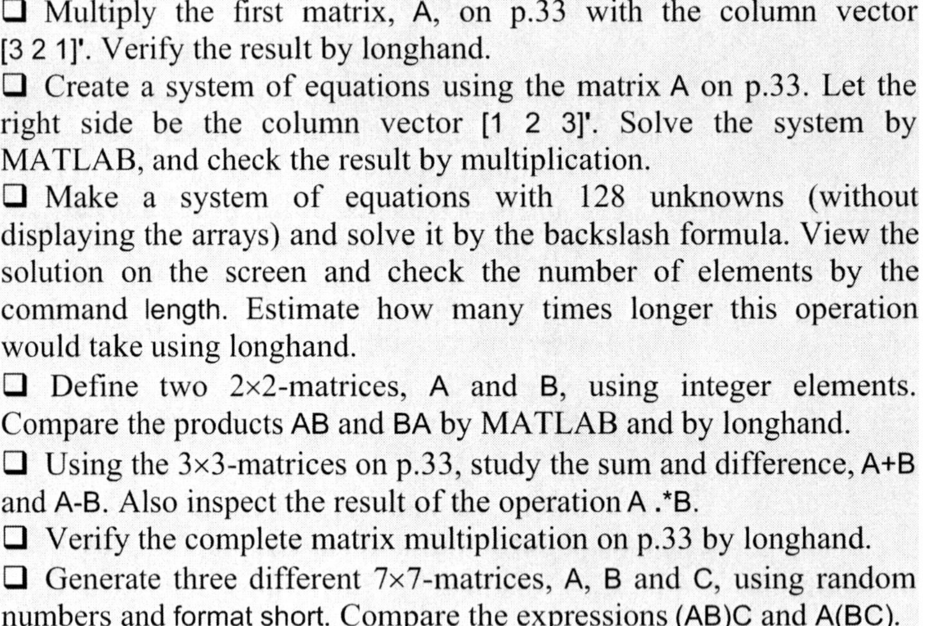

❑ Multiply the first matrix, A, on p.33 with the column vector [3 2 1]'. Verify the result by longhand.

❑ Create a system of equations using the matrix A on p.33. Let the right side be the column vector [1 2 3]'. Solve the system by MATLAB, and check the result by multiplication.

❑ Make a system of equations with 128 unknowns (without displaying the arrays) and solve it by the backslash formula. View the solution on the screen and check the number of elements by the command length. Estimate how many times longer this operation would take using longhand.

❑ Define two 2×2-matrices, A and B, using integer elements. Compare the products AB and BA by MATLAB and by longhand.

❑ Using the 3×3-matrices on p.33, study the sum and difference, A+B and A-B. Also inspect the result of the operation A .*B.

❑ Verify the complete matrix multiplication on p.33 by longhand.

❑ Generate three different 7×7-matrices, A, B and C, using random numbers and format short. Compare the expressions (AB)C and A(BC).

7 Plotting Functions

In order to represent a function $f(x)$ by a curve we need to compute $f(x)$ for a set of x-values. It is convenient to store the values x_i in a vector X. Fortunately, we need not type all these values for the square brackets. There is a simple way of generating equally spaced values of a variable x, as is evident from the following example.

```
X=0:0.5:10; Y=sin(X); figure(1), plot(X,Y), grid on
```

In the first statement we create a vector, containing elements starting at the value 0 with an increment of 0.5 up to a maximum value of 10. We choose to plot the function $y(x) = \sin(x)$, and by the second statement we compute the values of $y(x_i)$ corresponding to all x_i. The next step is to define the forthcoming *figure* as number 1, which makes it visible on top of other windows. After the actual plot command we add a *grid* of dotted lines at regular intervals.

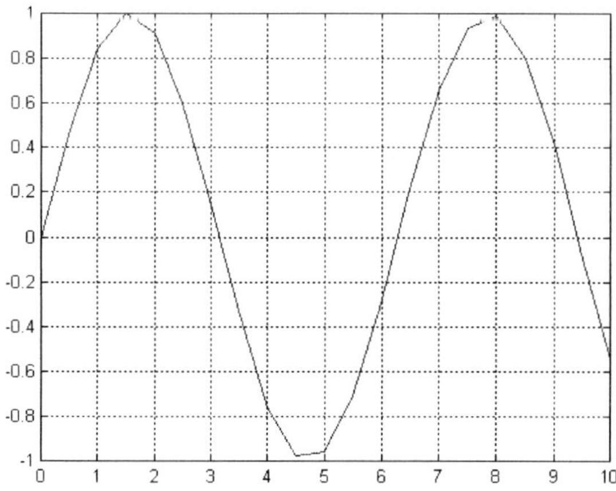

Evidently, this curve is not very impressive. As seen from the vector X, the interval between the points is as large as half a scale division, and that is the cause of the problem.

We did not obtain discrete (separated) points, however, since MATLAB has a built-in routine that automatically interpolates such data sets by linear segments. Instead of a rounded curve we thus obtain a polygon, connecting the sparse points defined by X and Y.

In order to obtain a smooth curve it is sufficient to shorten the intervals between the x-values. We try this by typing

```
X=0:0.02:10; Y=sin(X); figure(1), plot(X,Y), grid on
```

As demonstrated by the next figure, the curve now looks perfectly smooth.

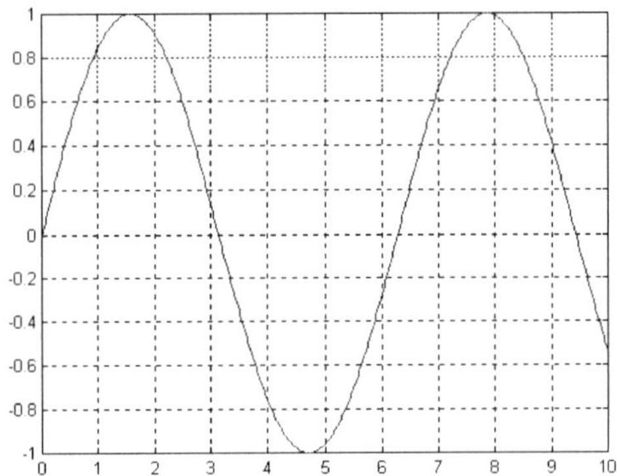

Next we turn to a slightly more complicated function, but the line required is similar.

```
X=0:0.02:20; Y=X.*cos(X); figure(1), plot(X,Y), grid on
```

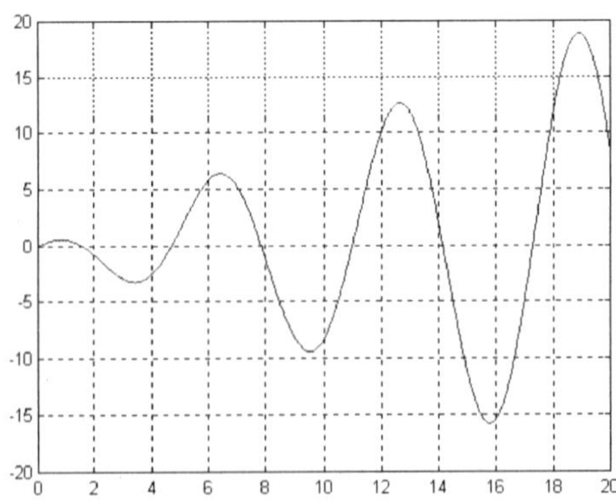

The above plot is what we obtain on executing the last command line.

The Help Command

As you go through this book you will occasionally encounter new commands and other details which may seem obscure to you. MATLAB provides on-line help to explain what the commands are for and also to describe the syntax, i.e. the correct way of using them. It usually includes related commands that may be useful for what you are attempting to do.

To obtain this assistance you only need to type, say,

```
help zoom
```

after the MATLAB prompt. For more general information, you may click on the word *Help* that you find on the top ribbon of the MATLAB Command Window.

Let us use the *help* information to zoom on the function $y = \sin(x^2)/x$ as follows.

```
X=0:0.01:20; Y=sin(X.^2)./X; figure(1), plot(X,Y), grid on, zoom on
```

Running this line we are warned about a division by zero. The first function value can't be calculated, since it is of the form $0/0$. The program simply omits this point, which most users find convenient. The next figure shows that the resulting curve oscillates with increasing frequency and diminishing amplitude towards large x.

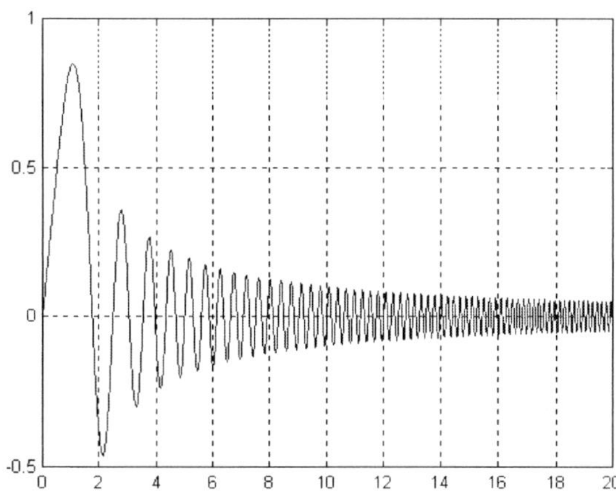

Having obtained this curve, we may enlarge part of it by pointing the mouse at a chosen point of the graph and by clicking on the left button. For instance, click repeatedly at the coordinate (18,0), and then return to the original figure by clicking a few times on the right button.

Plotting Two Functions Simultaneously

In many applications one wishes to compare different functions by showing them in the same diagram. To illustrate this, let us display the two branches of the function $y(x) = \sqrt{4 - x^2}$ by the lines

```
X=-2:0.01:2;  Y1=sqrt(4-X.^2);  Y2=- sqrt(4-X.^2);
figure(1),  plot(X,Y1,  X,Y2),  grid on
```

On the first line we defined two vectors, Y1 and Y2, one for the positive and one for the negative branch of the square root. On the second line we include the data for the two curves in the same plot command. These commands generate the following figure.

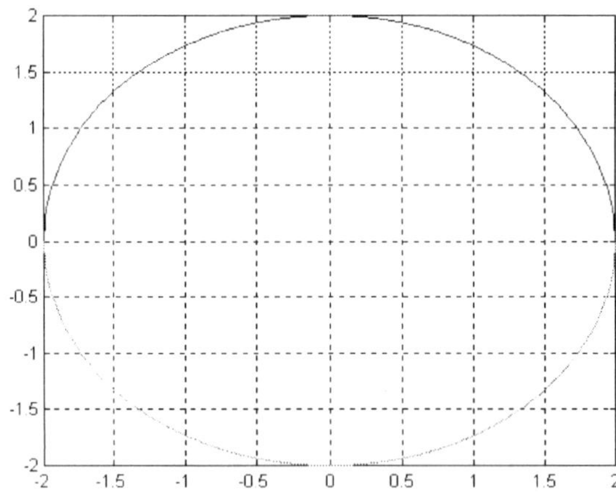

If you noticed that the function plotted is based on the equation for a circle, you may be astonished to see that the curve looks ellipsoidal. This is caused by the different size of the axes. We may correct this cosmetic problem by replacing the last line by

```
figure(1),  plot(X,Y1,  X,Y2),  grid on,  axis equal,  title('Circle')
```

At the same time, we add an appropriate title for the new figure. We do not have any more space available on this line, but we can still add labels for the axes afterwards (or just go on typing on the existing line).

```
figure(1), xlabel('x'), ylabel('y1, y2')
```

When you run this line the variable names immediately appear on the second plot.

Exercises

❑ Plot $f(x) = 1/(1+x^2)$ over a few suitable ranges of x.

❑ Plot $f(x) = e^{-3x^2}$ over a few suitable ranges.

❑ Plot the function $f(x) = x^{0.5} - x\sin(x)$ over a few ranges of positive x.

❑ Zoom on the preceding plot to locate the first two values of x where $f(x) = 0$.

❑ Plot the absolute value, $|f(x)|$, of the preceding function.

❑ Plot the polynomial $f(x) = x^3 - x^2 - x$ and search for zero crossings. Find the three solutions to the equation $x^3 - x^2 - x = 0$ numerically by zooming on these points.

❑ Write two lines to display the implicit function $y(x)$ defined by $x^n + y^n = 2^n$, where n=4. Plot the plus and minus branches of this expression separately in the same graph. Plot again with n=8 and n=64.

8 Script Files

In previous chapters we executed up to three MATLAB command lines in sequence, using results from a preceding line. We also recalled lines, modified them, and ran them again. Such juggling becomes inconvenient, however, in more complicated situations.

We have come to the point where we should use *script* files. This is a way of storing a set of commands on the hard disk, then executing them simply by typing the name of the file where they were stored.

Managing Script Files

A *script* file is a list of MATLAB command lines that you define for your own purposes. The family name (*extension*) of a script file is .M, which is the reason it is often called an *M-file*. While working with this book, you should store such files in a separate folder. The complete address of this folder, e.g. *c:\matlab\examples,* is known as the *path* to the file. To create such a folder, open the Windows *Explorer*, double-click on *c:\matlab*, select *File, New, Folder*, and then type *examples* over the highlighted area.

Before you use a script file for the first time you have to tell the program the path to the file. For this purpose, MATLAB has a special button (indicated below), named *Path Tool*[1].

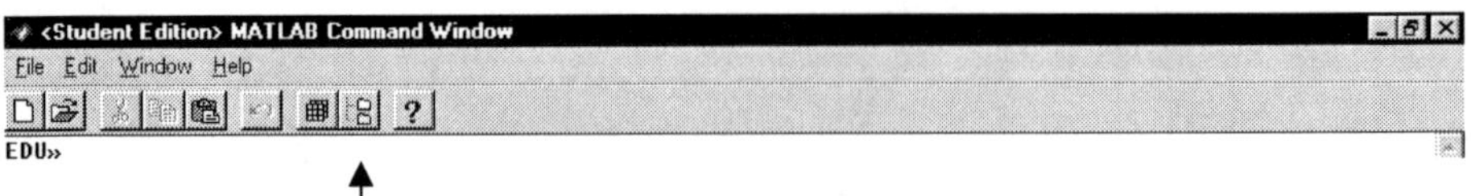

[1] If a message about MIPC appeared when you launched MATLAB, this means that you have not installed the *DialUp* software. If that is the case, you could insert the Windows CDROM and install the Communications part of the operating system. An alternative would be to declare your path by typing

```
path( path, 'c:\matlab\examples')
```

after the prompt every time you start MATLAB, or include it in your *Startup* file.

Just click on that button to obtain the following window.

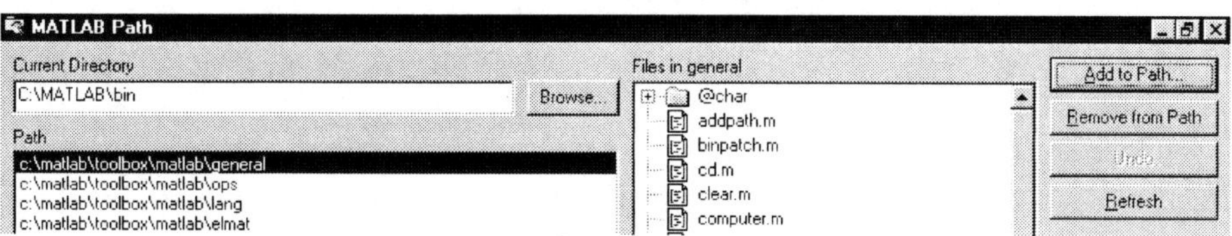

Next click on *Add to Path*, then double-click on the *matlab* folder, then on *examples*, and finally click on *OK* and on *Close*. Finally, you *save* these settings.

A Familiar List of Commands

Let us start by creating a file based on a simple example (p.10-11). From the MATLAB Command Window, select *File*, *New*, *M-file*. This opens the *Editor* window

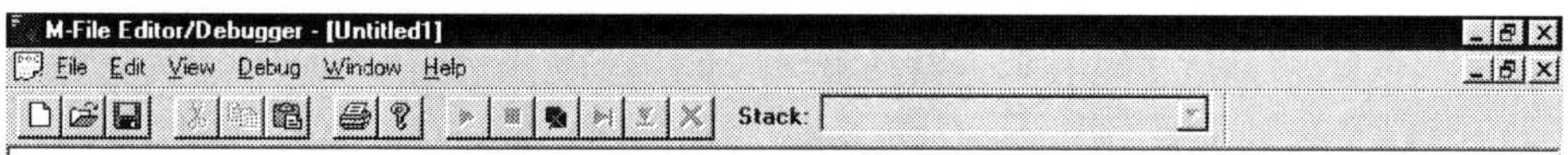

which is ready to receive your list of commands. For the first example, type the following text.

```
% ex81.m: Test of angles corresponding to sin(x)
format long
x=0.4,  s=sin(x)
% Alternative angles...
a1=asin(s),  a2=pi-a1,  a3=a1-2*pi,  a4=a1+2*pi
% Check these angles...
s1=sin(a1),  s2=sin(a2),  s3=sin(a3),  s4=sin(a4)
```

Notice that we have added segments beginning by %, which contain *comments* about the purpose of the file. These notes are only for the user, and the program ignores everything from the symbol % to the end of the line. As the first comment, we may enter the intended name of the file. Here, the first digit refers to the chapter number.

You save the above list by selecting *File*, *Save As*, browse for the path *c:\matlab\examples*, and enter the file name. It will not be necessary to add the extension *.m*, since this is taken for granted.

In many cases you will want to modify your file after running it, and thus it is convenient to keep the *Editor* running. Thus, do not leave by *Exit*, but simply click on some visible part of the MATLAB Command Window, or on the task button at the bottom.

In order to *run* the file, you select *File*, *Run Script*, enter the name of the file (without the extension *.m*), and push the *return* key. MATLAB then displays the results as if the commands had been typed line-by-line. The result is a long list of widely spaced lines containing variable names and values.

As an alternative, you may execute the file by entering its name (say, *ex81*) after the MATLAB prompt.

One advantage of using a script file is the overview it brings. In addition one can make the list easier to understand at a later time by adding comments. Finally, the file may serve as a starter for a longer sequence of calculations.

There is a dedicated button on the *Editor* window for opening an old file. In order to use a previous file as a template for a new one, you choose *File*, *Save As*, make the changes and finish by *Save* (third button at the top).

The text editor works the same way as most word processors as regards copying and pasting text fragments within a file, and between files.

Layout Preferences

On running the above file you find that MATLAB not only ignores the comments (which it should), but also omits them from the display. Another observation you might have made is that the program uses empty space very generously on the screen (several lines are used to display one value). Fortunately, we are allowed to set other preferences.

To eliminate empty space, you may select *File*, *Preferences*, *General*, *Compact*. In the same window you may put a check against *Echo On*, if you prefer to see your comments as well as the results. This has the effect of showing your commands too, however, which may be a disadvantage.

Whatever your choice, you can copy the results on the screen to your printer by *File*, *Print*. This act prints all the results from the beginning of the session, but you may also highlight a sequence of lines by the mouse and use *File*, *Print*, *Selection*.

An alternative mode, which gives you better control of margins and fonts is to use the *WordPad* program that comes with Windows. Just copy the highlighted segment by *Edit*, *Copy*, and then paste it into *WordPad*. It may be convenient to keep this program running during the entire MATLAB session.

Another Familiar Example

In the preceding example we added comments to explain the sequence of commands. The program did not display these comments along with the results, however. While we now revisit previous calculations concerning a triangle we shall also show how to add comments to the output. Let us repeat the analysis on p.14 by collecting the commands in a file, as follows.

```
% ex82.m: Triangle with Two Sides and One Angle Known
format long
disp('The two sides and the angle:')
a=4.1, b=9.1, A=20/180*pi
disp('Use the sine theorem to obtain sin(B), then B and C')
sB=b*sin(A)/a, B=asin(sB), C=pi-A-B
disp('Calculate the side c by the sine theorem')
c=a*sin(C)/sin(A)
disp('Convert the angles to degrees')
Adg=A/pi*180, Bdg=B/pi*180, Cdg=C/pi*180
disp('Repeat to find the second solution')
a=4.1, b=9.1, A=20/180*pi
sB=b*sin(A)/a, B=pi-asin(sB), C=pi-A-B
disp('Convert the angles to degrees')
Adg=A/pi*180, Bdg=B/pi*180, Cdg=C/pi*180
```

Here, we have used the command *disp* to display the comments within quotes, and these comments will be visible on the screen, as well as serving as reminders in the file. The option echo on would duplicate these comments, however, so if you have chosen the echo, this option should now be unchecked.

The beginning of the output display now looks like this.

```
The two sides and the angle:
a =
   4.10000000000000
b =
   9.10000000000000
A =
   0.34906585039887
Use the sine theorem to obtain sin(B), then B and C.
sB =
   0.75911787908868
B =
   0.86195692026160
C =
   1.93056988292933
Calculate the side c by the sine theorem.
c =
  11.22011089690112
Convert the angles to degrees.
Adg =
    20
Bdg =
  49.38649365308416
Cdg =
    1.106135063469158e+002
```

This presentation definitely is more readable, although one would wish that a variable name and the value were given on a single line. We might achieve this by detailed formatting of each output item, but that is hardly worth the trouble.

A System of Linear Equations

There is nothing to stop us from making a script file that solves a system of equations and compares left and right members after the solution. The file below illustrates this.

```
% ex83.m: A System of Five Unknowns
echo on
format short,  clear,  A=rand(5,5)  % Random, square matrix
B=rand(5,1)                         % Random column vector
X=A\B                               % Solve for X
format long,  left=A*X,  right=B    % Compare left and right members
format short,  error=left-right     % Show the difference
meanB=sum( abs(B))/5                % Mean value of B for comparison
```

We have seen that there are several advantages to a script file. It brings a permanent record of successful calculations, which can be used as a model for other tasks. In addition, the comments we supply when typing the file make it easier to understand the purpose of the command lines. Finally, such comments may also be transferred to the list of results.

Startup File

We can also use an M-file to make MATLAB settings permanent. As the program starts, it looks for a file named *startup.m* and runs it automatically. The *Preferences* under *File* are examples of what such a file may contain, but you may include any MATLAB command that would serve your purposes.

To illustrate the use of a *startup* file, let us set the output format to be *compact* and specify that we want all commands and comments echoed when the results are displayed. Finally we specify the path where the *Editor* should look for your M-files, a path to be modified to meet your current situation.

```
% startup.m: Initialize by these Commands
format compact
echo on                                  % ... if you prefer that
format long                              % ... if you prefer that
cd c:\matlab\examples                    % Modify to fit your current path
```

When we have typed this file, we must store it in the same folder as the MATLAB program, which normally is *c:\matlab\bin*. This means that we do not have to specify this path, since it is MATLABs own. Once this file is installed, we may run it if we wish, but later the program will take care of that.

Exercises

❑ Create a script file that solves for sides and angles as on p.13. Include suitable comments.

❑ Expand *ex82* to calculate the area of the triangle for the two alternatives, and in two different ways.
❑ Make a script file to solve for sides and angles of the triangle on p.20. Include a check by summing the angles.
❑ Write a script file to calculate the volume and the surface of the parallelepiped on p.22.
❑ Using a script file, plot the functions $\sin(x)$ and $\cos(x)$ in the same diagram over the interval $0 < x < 3\pi$. Include a title and labels for the axes. Then try to add a third curve for $\tan(x)$.

9 Functions and Function Files

In this chapter we shall plot some frequently used functions and study some of their properties, while learning about new means for plotting.

Let us start with the basic trigonometric functions, by displaying all of them in the same diagram. For this we need the following script file.

```
% ex91.m: Plot Trigonometric Functions
echo off,  x0=5;  X=-x0:0.01:x0;                % Range for calculations
Y1=sin(X);  Y2=cos(X);  Y3=tan(X);  Y4=cot(X);
figure(1),  plot(X,Y1,  X,Y2,  X,Y3,  X,Y4),  grid on
y0=3;                                            % Vertical plot limit
axis([-x0 x0 -y0 y0])                            % Ranges of the plot box
title('Trigonometric Functions'),  xlabel('x'),  ylabel('sin(x), cos(x), etc.')
legend('sin(x)', 'cos(x)', 'tan(x)', 'cot(x)')          % Identify curves
```

In planning this file, we anticipate that $\tan(x)$ and $\cot(x)$ will take arbitrarily large values close to certain values of x. Hence, we cut off the plot to display only function values in the range $-y_0 \leq y \leq y_0$. We achieve this by the axis command. The result of this script file is shown below.

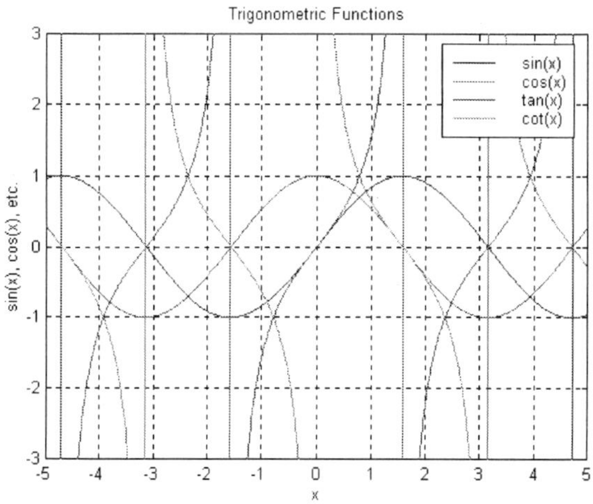

In the above plot we like to identify the four different functions by their colors. The legend command does this for us, if we supply a description of all the curves, in the order they occur in the plot parentheses.

You may have wondered why there are vertical lines in the plot at points where some function crosses zero. These are caused by the linear interpolation inherent in the program. In the region of x where $\tan(x)$ crosses zero $\cot(x)$ flips from minus to plus infinity. In practice, this means that the plot routine draws a straight line from the lowest value calculated to the highest one. A similar explanation holds for the zeros of $\cot(x)$.

Inverse Trigonometric Functions

We have already used some inverse functions, such as $\arcsin(x)$ on p.10. Let us now plot the inverse functions corresponding to those in the preceding section. Just modify *ex91.m* to read

```
% ex92.m: Inverse Trigonometric Functions
x0=1;  X=-x0:0.01:x0;              % Range for calculations
Y1=acos(X);  Y2=atan(X);
figure(1),  plot(X,Y1, X,Y2),  grid on
y0=2*pi;                                    % Vertical plot limit
axis([-x0 x0 -y0 y0])                       % Ranges of the plot box
title('Inverse Functions'),  xlabel('x'),  ylabel('acos(x), atan(x)')
legend('acos(x)', 'atan(x)')                % Identify curves
```

From the resulting plot (below) we find that both functions are plotted as single-valued, and $\arctan(x)$ becomes almost a straight line over the range chosen. Evidently, MATLAB plots *fundamental* function curves, and we must imagine them shifted up or down by multiples of 2π to complete the presentation.

We can devise another way of plotting inverse functions, however. If $y = \cos(x)$, say, then $x = \arccos(y)$, and a plot of $\cos(x)$ hence contains all the x- and y-values we need to construct a plot of $\arccos(x)$.

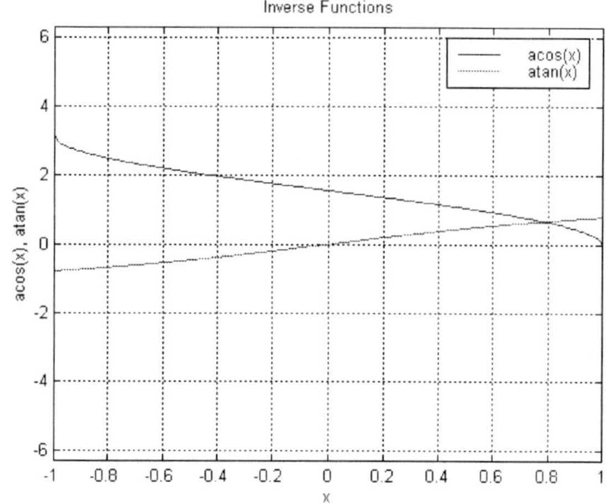

To obtain several branches of arccos(x) and arctan(x) it is enough to exchange the vectors X and Y as follows, using *ex92.m* as a template.

```
% ex92a.m: Inverse Trigonometric Functions by Swapping
x0=2*pi;  X=-x0:0.01:x0;         % Range for calculations
Y1=cos(X);  Y2=tan(X);
figure(1),  plot(Y1,X, Y2,X),  grid on
y0=2*pi;                          % Vertical plot limit
axis([-x0 x0 -y0 y0])             % Ranges of the plot box
title('Inverse Functions'),  xlabel('x'),  ylabel('acos(x), atan(x)')
legend('acos(x)', 'atan(x)')      % Identify curves
```

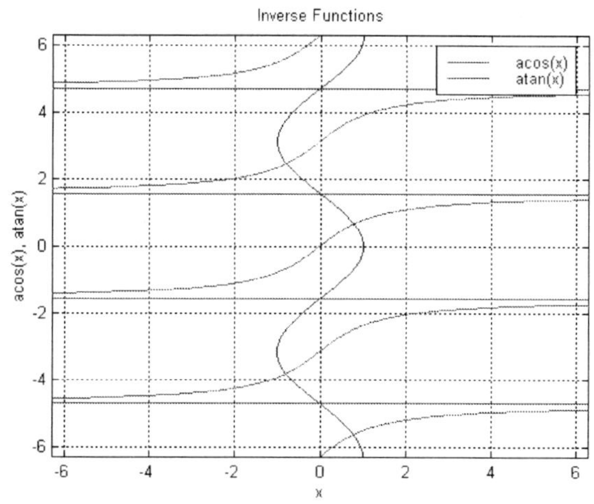

Exponential Functions

The basic exponential function, e^x, may also be written $\exp(x)$. This function exists for both positive and negative values of x, and we could plot it by the following file.

```
% ex93.m: Plot Exponential Functions
x0=10;  X=-x0:0.01:x0;
Y1=exp(X);  Y2=1e3*exp(X);  Y3=1e6*exp(X);
figure(1),  plot(X,Y1, X,Y2, X,Y3),  grid on
title('Exponential functions'),  xlabel('x'),  ylabel('exp(x), etc.')
legend('exp(x)', '1e3*exp(x)', '1e6*exp(x)')
axis([-x0 x0 0 2.5e4])              % Ranges of the plot box
```

Here, we study three functions in the same plot, differing only in the multiplying constant. The axis command confines all the curves in the same range of y-values, in spite of the large constant factors. It is evident from the figure below that the curves are similar, but shifted in the horizontal direction. We may understand this by rewriting the multiplying constant in exponential form.

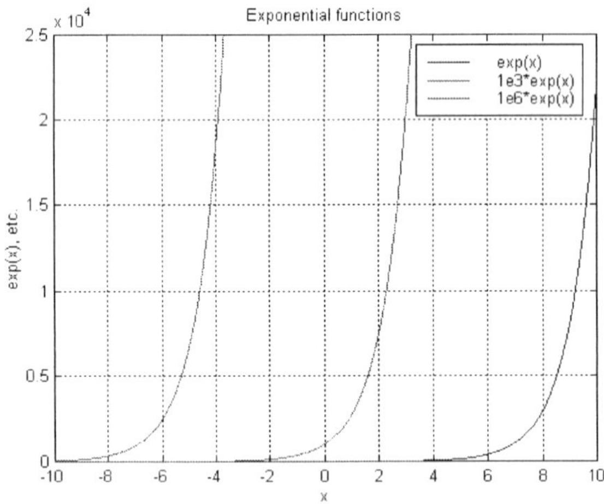

It is clear from the above that a plot of the general exponential function, a^x, will have same appearance.

Logarithmic Functions

The inverse function to $y = e^x$ is of course the logarithmic function $x = \ln(y)$. We may plot $\ln(x)$ by the following file, similar to *ex93*.

```
% ex94.m: Plot  Logarithmic Functions
x0=10;  X=0.01:0.01:x0;
Y1=log(X);  Y2=log(1e3*X);  Y3=log(1e6*X);
figure(1),  plot(X,Y1, X,Y2, X,Y3),  grid on
title('Logarithmic functions'),  xlabel('x'),  ylabel('log(x), etc.')
legend('ln(x)', 'ln(1e3*x)', 'ln(1e6*x)')
```

It is clear from the plot (below) that multiplying the argument (x) by a constant factor shifts the curve in the vertical direction. Obviously, this figure is similar to the preceding one, rotated by 90°. In fact, we may also obtain these curves by plotting exponential functions with X and Y exchanged.

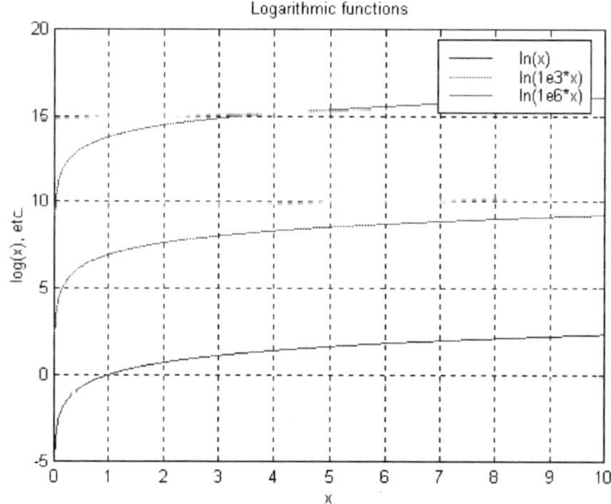

In a similar way, we may plot $\log(x)$ using the MATLAB notation log10(x), or $\log_2(x)$ as log2(x), which yields curves of the same shape.

Logarithmic Plots

When a function varies strongly over one region of x or y, and little over another, it may be advantageous to use non-linear scales on the axes. Logarithmic scales are readily available in MATLAB, as is evident from the next example.

```
% ex95.m: Plot Exponential Functions
x0=100;  X=-x0:0.1:x0;
Y=exp(X);
figure(1),  semilogy(X,Y),  grid on
title('Exponential function'),  xlabel('x'),  ylabel('exp(x)')
```

The resulting curve is a trivial straight line covering the range $10^{-50} < y < 10^{50}$. Let us now modify this file to display the more complicated functions $y_1 = \exp(-x^2)$ and $y_2 = \exp(-3x^2)$.

```
% ex95a.m: Plot Gaussian Functions
x0=5;  X=-x0:0.01:x0;
Y1=exp(-X.^2);  Y2=exp(-3*X.^2);
figure(1),  semilogy(X,Y1,  X,Y2),  grid on
title('Gaussian functions'),  xlabel('x'),  ylabel('exp(-x^2), exp(-3x^2)')
legend('exp(-x^2)',  'exp(-3x^2)')
```

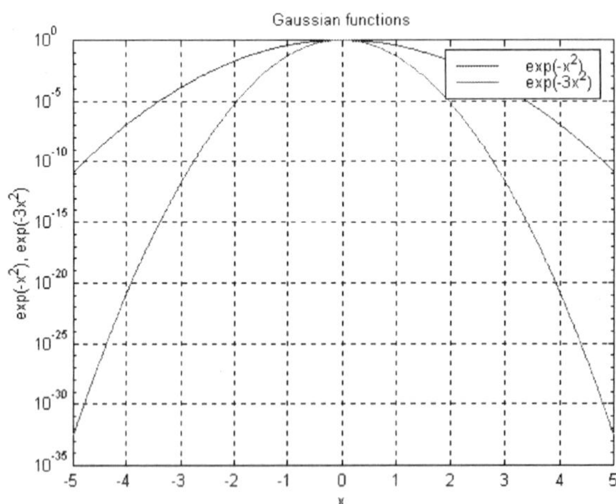

The above two curves evidently show the variation of the Gaussian functions even in the regions where the values are very small.

Inline Functions

MATLAB contains a number of built-in elementary functions. We may of course combine these to create a composite function. For a simple function that can be defined by one statement, it is convenient to attach a name to it, so that we can use the function with any argument, as in the case of $\sin(x)$. This type of object is called an *inline* function.

As an example of an inline function we take

```
gauss=inline( 'exp(-x^2)' )
```

Once this command has been executed, we may use gauss(x) to obtain single values of the function, such as

```
g0=gauss(0),  g1=gauss(1)
```

In this case, we assumed x to be scalar, but we are not restricted to this usage. When we employ an uncommon variable name, say X, to denote a vector we must declare it as the independent variable by a separate character within quotes, as in the example

```
gauss=inline( 'exp(-X.^2)', 'X' );  X=0:3,  gauss(X)
```

which gives us four values of the Gaussian function.

An inline function may also take more than one argument. For instance, we could write a general definition for the Gaussian as follows.

```
gaussian=inline('exp(-X.^2/a^2)', 'X', 'a');
```

We demonstrate how to use this function by the following script file.

```
% ex96.m: Plot Gaussian
close all                              % Erase elements of earlier plots
X=-3:0.01:3;  Y=gaussian(X,1);
figure(1),  plot(X,Y),  grid on,  title('Gaussian')
```

Here, we have added the command *close all*, which removes all previous plots, in particular an intrusive legend. The resulting plot is familiar.

To display two different curves in the same plot, we need to type a new script file, i.e.

```
% ex96a.m: Plot  two Gaussians
X=-3:0.01:3;  Y1=gaussian(X,1);  Y2=gaussian(X,0.2);
figure(1),  plot(X,Y1,  X,Y2,'--'),  grid on,  title('Gaussians')
legend('a=1.0', 'a=0.2')
```

The following figure shows plots for two values of the parameter. As before, we use legend to identify the curves, here by line type (dashed).

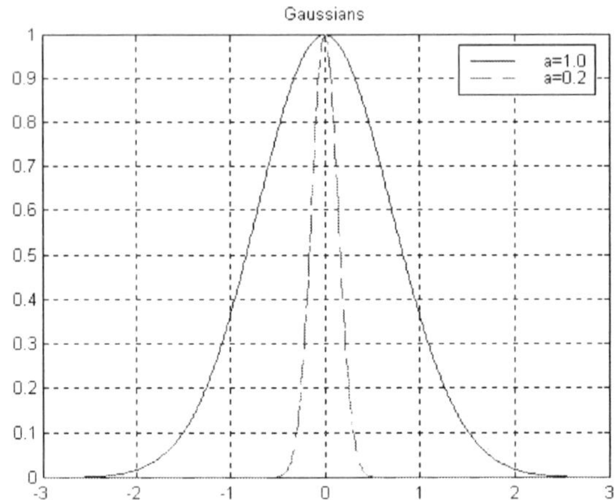

Function Files

An inline function is stored in primary memory and only remains valid during the MATLAB session where it was defined. We shall see, in a later chapter, that the Gaussian is an important function in an uncertain world. It is the most frequently used function in statistics, and it is convenient to have it ready-made, for use in future script files. To achieve this we create the function file below.

```
% gaussian.m: Gaussian Function with a Parameter
function  F= gaussian(X,a);
F=exp(-X.^2/a^2);
```

After saving this file, you will be able to call the function gaussian from any script file. You may immediately try it with a vector argument directly from the MATLAB window.

The difference between a *script* file and a *function* file is that the latter accepts *arguments* in parentheses after the name. The *function* declaration should stand alone on its line, since the program does not continue to read beyond the declaration statement. We also wish the file to be "silent", so we close all statements by a semicolon to inhibit display of the results.

We generally are free to name M-files as we please, but function files are different. If the file and the function should get different names, MATLAB assumes the *file* name as the name for the function. Thus, to avoid confusion the names should be the same.

The result of the command

```
X=0:0.5:3,  Y=gaussian(X,1)
```

illustrates that a row vector X yields a row vector Y containing the function values. If we prefer a column output, we may easily arrange that by typing

```
Y=gaussian(X',1)
```

Function File Yielding Two Output Vectors

So far we have plotted curves where y was expressed explicitly as a function $y = f(x)$. A curve may also be described by two separate functions

$$\begin{cases} x = f(t) \\ y = g(t) \end{cases}$$

where the *parameter* t is understood to run through a range of values. As a simple example of this *parametric* definition, let us take

$$\begin{cases} x = a_x \cos(t) \\ y = a_y \cos(t - t_0) \end{cases}$$

which we implement by the following function file. Earlier, we had functions yielding a vector as the output, but MATLAB has little restrictions in this regard. Here, the output object may be seen as a vector containing two vectors. The argument T is an ordinary vector containing the values of the parameter t.

```
% ellipse.m: Function Defining an Ellipse
function [X,Y]=ellipse(T, ax, ay, t0);        % Comma between X and Y
X=ax*cos(T);
Y=ay*cos(T-t0);
```

Notice that a *comma* between X and Y is required in a function *definition*. Let us now test this function with different arguments by the following script file, which largely speaks for itself. The plot now must handle three pairs of vectors.

```
% ex97.m: Plot Ellipses
T=0:1e-3:2*pi;                          % Values for common variable
[X1 Y1]=ellipse(T, 2, 1, 0);
[X2 Y2]=ellipse(T, 2, 1, pi/3);
[X3 Y3]=ellipse(T, 2, 1, pi/2);
figure(1), plot(X1,Y1, X2,Y2, X3,Y3), grid on, axis equal
legend('t0=0', 't0=pi/3', 't0=pi/2')
```

In the figure below, we have dragged the legend to the left by the mouse. The first curve is trivial, because the *x*- and *y*-values are strictly proportional to each other, which only produces a straight line. The other curves look much like ellipses of different orientation.

If we expand the cosine in $y = a_y \cos(t - t_0)$, we may eliminate the factors involving t to obtain a relation between x and y. The latter turns out to be a 2nd-order equation, and since the excursions in x and y are limited, this curve can only be an ellipse.

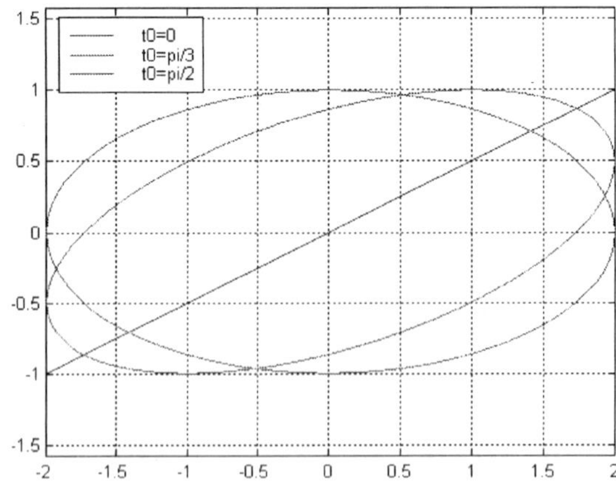

Curves in Polar Coordinates

Some curves are conveniently defined in polar coordinates (below) by the function $r(\theta)$. Converting to (x, y) coordinates we have

$$\begin{cases} x(\theta) = r(\theta)\cos(\theta) \\ y(\theta) = r(\theta)\sin(\theta) \end{cases}$$

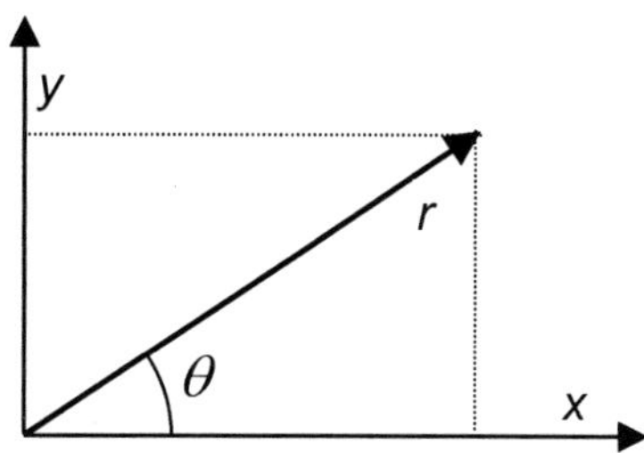

We only need to modify *ellipse* slightly to illustrate the usage, keeping T as the notation for the common vector of angles θ. The vector R contains values of $r(\theta)$.

```
% rosette.m:  Function Defined in Polar Coordinates
function [X,Y]=rosette(T, n);
R=cos(n*T);
X=R.*cos(T);  Y=R.*sin(T);
```

The corresponding script file plots two sets of vectors over a full turn of θ. The resulting figure is shown below.

```
% ex98.m: Plot Rosettes
T= 0:1e-3:2*pi;
[X1 Y1]=rosette(T, 1);
[X2 Y2]=rosette(T, 2);
figure(1),  plot(X1,Y1,  X2,Y2,'--'),  grid on,  axis equal;
  title('Rosettes'),  legend('R=cos(T)', 'R=cos(2*T)')
```

A polar plot is of course a special case of the parametric one (p.57). The latter permits us to specify any curve in the (x,y) plane. For instance, it permits us to define a loop not enclosing the origin, whereas polar coordinates would require $r(\theta)$ to possess more than one branch.

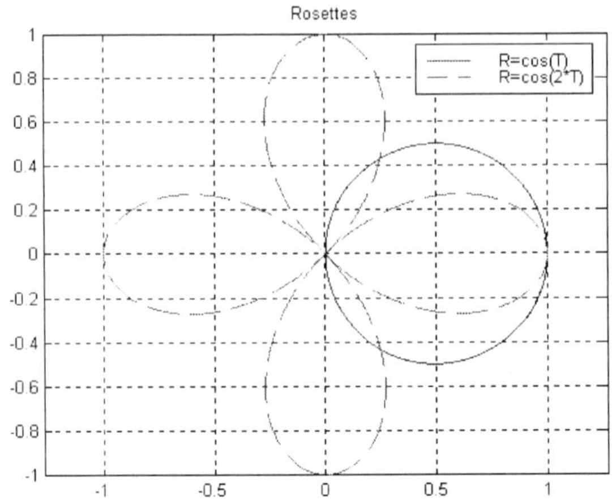

Exercises

❑ Modify *ex91* to plot sin(20x)+sin(x), after removing axis and legend commands. Then plot sin(20x)sin(x).

❑ Using *ex91* as a template, determine the number of elements in X and plot a random function. Then plot 10cos(x) + (random). Finish by plotting cos(x)(random).

❑ Modify *ex92* to plot ln(x), $\log_{10}(x)$, and $\log_2(x)$ in the same graph over a suitable range of x. Remove the axis command.

❑ Repeat *ex93* with other choices of the y range. Also try removing the axis command.

❑ Modify *ex96a* and its function files to plot the Gaussian function for three parameter values, a=0.1, 0.3, 1.0, in a diagram with a logarithmic y-scale. Expand the x range by a factor of 10.

❑ Try other values of ax, ay, and t0 to explore the types of curves that may be obtained with *ex97*.

❑ Using *ex98* as a template, try the radial functions sin(0.5θ) and sin(1.5θ). Also plot r against θ separately (figure(2))

❑ Predict what will happen to the curves, if the functions cos in *ex98* are replaced by sin. Then plot the new file.

10 Zeros and Extreme Points

Solving the equation $f(x) = 0$ amounts to finding the zero crossings of the function. It is customary to use the short name *zero* for an x-value that makes the function zero. We may detect zeros manually by plotting and zooming, but we shall now make this search automatic.

Let us start with the simple polynomial

$$f(x) = x^3 - 2x^2 - x + 2$$

If we manage to find the zeros of this function, we have also solved the algebraic equation

$$x^3 - 2x^2 - x + 2 = 0$$

Simplified Plotting

We begin by plotting the function to obtain the approximate location of the zeros. When we have defined a function by an inline statement or a function file, a simplified plot procedure, *fplot*, becomes available (see *help fplot*). We need not use the vector form for the function, and we specify the plot range by a vector as the second argument.

```
% ex101.m: Plot a Cubic Polynomial
f=inline('x^3- 2*x^2- x+2');                    % Define function f(x)
figure(1), fplot(f, [-2 3]), grid on, zoom on
  title('x^3-2*x^2-x+2'), xlabel('x')
```

Inspecting the figure below we see that there are zeros at about –1, 1, and 2, and by clicking several times on the first of these crossings we are able to pin down that value to just above –1, say -0.99999. The results for the other zeros are similarly close to an integer, and you may already have suspected (and verified) that all three zeros are exact integers. In fact, we have chosen this case to be able to check the accuracy of our numerical estimates. Before leaving the process of zooming we note that the accuracy of the zeros would be sufficient for most practical purposes.

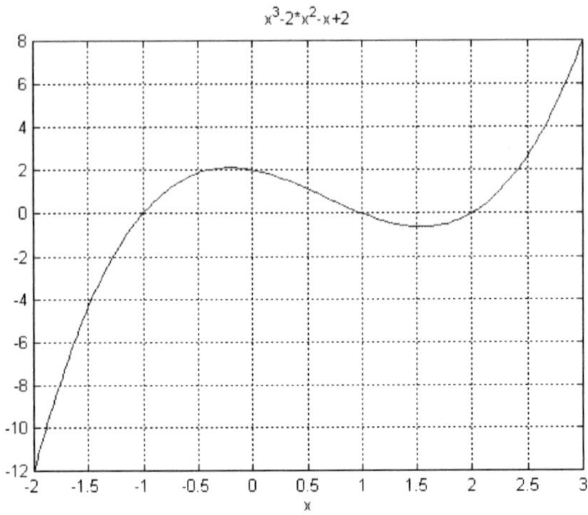

Solving by Bisection

We now come to an alternative method for finding zeros. The following figure shows the above function $f(x)$ over a smaller interval. We start by specifying two values, x_1 and x_2 (chosen to be larger than x_1), that bracket the zero crossing.

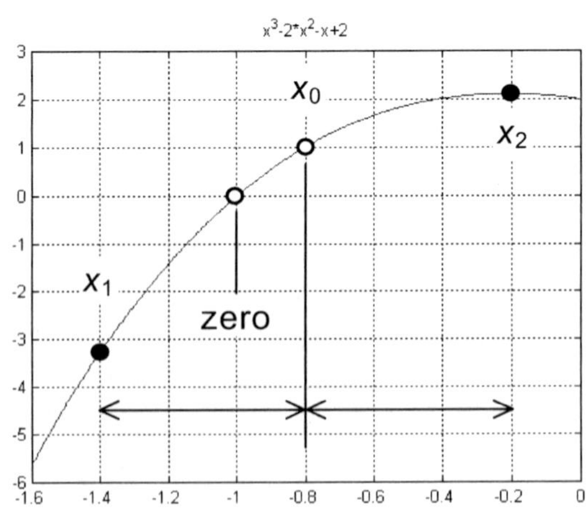

Our strategy for finding a zero could not be simpler. We just divide the interval (x_1,x_2) into two equal parts to obtain the trial value x_0. After calculating $f(x)$ for that coordinate, we decide whether the result is below zero or above. If it is above zero, we discard the value x_2 and replace it by x_0. If it falls below zero, we instead replace x_1 by x_0. By this simple recipe we have thus halved the interval in which the zero is located. If we repeat the procedure many times, the value x_0 should become a good approximation for the zero crossing. The limitation is only the number of digits handled by the program. We would perhaps not attempt to do this by hand, but the computer relieves us of the expected tedium.

Solving a Cubic Polynomial Equation

Let us now use the bisection scheme to find zeros for the polynomial we plotted in the beginning of this chapter. The following script file is designed to extract a zero for this function. The initial values x_1 and x_2 need not be very close to the solution, as long as they bracket only one of the zeros. As a first step we use the *inline* procedure to define the function under investigation.

The section starting by *while* and finishing by *end* is called a *loop*, since it is designed to repeat itself. In this loop we calculate x_0 according to the above principle and replace one of the end coordinates by this new value. The command *pause* permits us to inspect the results of each turn of the loop as the calculations proceed.

```
% ex102.m:  Find Zeros by Bisection
clear,  format long,  echo on
f=inline('x^3- 2*x^2- x+2');                          % Define function f(x)
figure(1),  fplot(f, [-2 3]),  grid on,  zoom on      % Plot survey
x1=-1.4;  x2=-0.2;                 % Initial values, x2>x1
while 1==1                         % Repeat forever
  f1=f(x1);  f2=f(x2);
  x0=(x1+x2)/2;  f0=f(x0);
  x1_x0_x2=[x1 x0 x2]              % Equidistant coordinates
  f1_f0_f2=[f1 f0 f2]              % Display function values
  if f0*f1>0,  x1=x0;  else x2=x0;  end              % Replace one value
  error=abs(x2-x1)                 % Safe error estimate
```

```
  if error<1e-15, break, end        % Stop the loop
  pause                             % Push <Return> to continue
end                                 % Endpoint of loop
x0                                  % Display final value
```

The main principle of the procedure lies in the if...else...end clause. There we test if f0*f1 is above zero, i.e. if these function values are of the same sign. If this is the case, we make the replacement x1=x0 and otherwise x2=x0. These statements are nearly in plain English.

Here, we have specified that the loop is to repeat itself while 1==1. The double equal sign indicates that 1 is *compared* to 1, and since 1 is always equal to 1, the loop turns forever. We could also have used the shorter expression while 1 (1 meaning "true"). We break this loop, however, when the end values x_1 and x_2 have become nearly indistinguishable.

Running this file we may study how the strategy is applied in a loop, and then push *return* for the next turn of the loop. Gradually the difference between x1 and x2 vanishes. After some fifty turns the file finishes and renders the final value. You may easily verify that this is the expected number of bisections required to attain the final accuracy.

A More Flexible Script File

From the plot on p.62 we saw that the above polynomial has three zeros, but we only calculated one of them. In the following script file we modify *ex102* to let us enter the values of *x*1 and *x*2 from the keyboard by *input* statements. This permits us to bracket any zero we wish to compute. We also remove the display and *pause* commands to make the file run to the end in a short time.

```
% ex103.m: Find Zeros by Bisection
clear; format long; echo off
f=inline('x^3- 2*x^2- x+2');                          % Define function f(x)
figure(1), fplot(f, [-2 3]), grid on, zoom on         % Plot survey
x1=input('x1= '); x2=input('x2= ');                   % Bracketing values
while 1==1                                            % Repeat forever
  f1=f(x1); f2=f(x2);
  x0=(x1+x2)/2; f0=f(x0);
```

```
  if f0*f1>0,  x1=x0;  else x2=x0;  end              % Replace one value
  if abs(x2-x1)<1e-15,  break,  end                   % Stop the loop
end                                                   % Endpoint of loop
x0                                                    % Display final value
```

Here, the *input* command first displays a string as a prompt for the user. Then it accepts a value from the keyboard. Using this new file, it is easy to bracket the remaining zeros by x1 and x2 and to calculate them.

The Built-in Algorithm

MATLAB provides the built-in function *fzero* for finding zeros of a function. We may run it directly from the Command Window by the line

```
f=inline('x^3- 2*x^2- x+2');  x0=fzero( f, [0 1.5], 1e-15)
```

The second argument of *fzero* is a vector containing the bracketing values x1 and x2, and the third argument sets the tolerated error for x0.

We have seen that these numerical procedures let us calculate roots of a cubic algebraic equation (or zeros of a polynomial) with an accuracy of about 15 figures, which we could hardly achieve by plotting and zooming.

It is well known that algebraic equations of the 4th degree or lower may be solved in terms of elementary functions. The exact roots may be extremely complicated, however, and it is often necessary to write long programs to obtain numeric values. The procedures that we have just tested let us find all real roots of any algebraic equation.

Transcendental Equations

Equations do not always consist of algebraic terms. For instance,

$$\cos(x) - x = 0$$

is known as a transcendental equation and has no exact solution that can be expressed in standard functions. It is as easy to solve numerically, however, as any algebraic equation.

Before solving it, we would like to have an overall view of the function on the left side.

```
f=inline('cos(x)-x'); figure(1), fplot( f ,[-4 4 -2 2]), grid on
```

The first argument in fplot consists of the function defined by the *inline* statement. The second one is the vector $[x_{min}\ x_{max}\ y_{min}\ y_{max}]$. The figure below shows the resulting plot, which includes only one zero. If we consider the functions $\cos(x)$ and x separately, we realize that they can only meet at one point. We just make another plot by

```
f=inline('cos(x)'); figure(2), fplot( f ,[-4 4 -2 2]), grid on
```

and imagine the straight line corresponding to $y = x$.

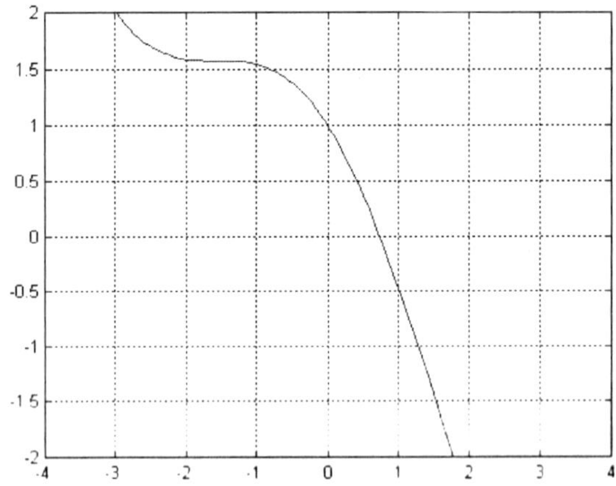

We calculate the root by the command line

```
f=inline('cos(x)-x'); format long, x0=fzero(f, [0 2], 1e-15)
```

to obtain x0=0.73908513321516.

Verifying Solutions

In particular when we use *fzero*, the solution appears as if by magic and you might be skeptical about its accuracy. No matter what method you use to solve an equation you may test if the value really makes the left member of the equation vanish. As is usual with numeric calculations, however, we can't expect an exact zero. Having obtained a solution x_0, we may test the left member with the values $x_0 - \varepsilon$ and $x_0 + \varepsilon$, the increment ε being the expected error. One of these values should make the left member change sign.

Let us apply this test to the preceding example. We first solve as before to obtain x0. Then we execute the following line.

```
f=inline('cos(x)-x'); format short, f(x0-1e-15), f(x0+1e-15)
```

The results for *f* are 1.6653e-15 and -1.7764e-015, which shows that the error in x0 is indeed smaller than 1e-15. The function values are almost symmetrical about zero, and this suggests that the solution might be further improved by interpolation.

Minimum and Maximum Points

The plot on p.62 demonstrates that the function has one maximum and one minimum. These points are hard to determine by zooming, because the slope of the curve vanishes there. We could expect to find such points numerically, however, by a bisection process similar to what we used for zeros.

In fact, MATLAB has a built-in function for finding the lowest value in a given interval. We may test this procedure by the line

```
f=inline('x^3- 2*x^2- x+2'); xm=fmin(f, 1, 3, 1e-8), fxm=f(xm)
```

The second and third arguments in *fmin* specify the limits of the interval where we search for the minimum value. The last argument is the requested accuracy for the answer. By the above line we find an accurate value of xm and the corresponding value of f.

If we now explore an interval at the left end of the scale

```
f=inline('x^3- 2*x^2- x+2'); xm=fmin(f, -2, 0, 1e-8), fxm=f(xm)
```

we find that *xm* becomes equal to *x* at the extreme left. This reminds us that the lowest value need not be in the interior of the interval and need not be associated with a horizontal tangent.

The curve on p.62 also exhibits a *maximum* at about x=-0.25. How do we evaluate the position *xm* of this point? By changing the sign of the function we transform a maximum into a minimum, and we may then use the same formalism as before. We only need to remember to restore the sign of the function value, f(xm).

```
f=inline('-(x^3- 2*x^2- x+2)'); figure(1), fplot(f,[-2 3 -12 8]), grid on
xm=fmin(f, -1, 1, 1e-8), fxm=-f(xm)
```

Exercises

❑ Solve the equation $x^2 - 3x - 3 = 0$ completely using *ex103*. Just change the *inline* definition and use trial and error to find a suitable range for *fplot*.

❑ Solve $x^5 - 15x^4 + 85x^3 - 225x^2 + 274x - 119 = 0$ completely by a modified version of *ex103*.

❑ Find all zeros of the function $f(x) = \sin(x) - 0.5x$ by a modified version of *ex103*.

❑ Solve $\tan(x) - \sqrt{x} = 0$ over the interval $0 < x < 10$ by means of *fzero*.

11 Symbolic Algebra

We are now entering a new field of computing, where results appear in the form of symbolic expressions, rather than numeric values. If you have the Student Edition or Student Version (US), this facility is available. By the use of familiar commands this toolbox permits access to a subset of Maple®.

Symbolic algebra is somewhat related to word processing. You enter an expression in the form of ordinary characters, and a program transforms it according to the well-established rules of mathematics. These rules tell us how to solve equations by adding equal terms to both sides, multiplying all terms by the same expression, and so on. The main difficulty for human beings is to manipulate expressions repeatedly without making mistakes. The computer never loses concentration during long and boring transformations, and it is thus eminently suited for this type of work. In this chapter we shall have a quick look at the main features of symbolic algebra.

Solving Algebraic Equations

Symbolic algebra yields *exact* results. To enter this mode of operation, we must declare the variables as *symbolic* by the command *syms*. Let us start by solving the cubic equation from the preceding chapter. It is sufficient to type the following line, containing the left member of the equation.

```
clear, syms x, x0=solve(x^3-2*x^2-x+2)
```

MATLAB responds by giving the three roots in the form of the column vector

```
x0 =
[  1]
[  2]
[ -1]
```

where we recognize the approximate solution obtained earlier.

This also works with a more complicated equation where the roots are exact integers. We need not repeat the *syms* command.

```
x0=solve(x^5-15*x^4+85*x^3-225*x^2+274*x-120)
```

Recasting Expressions

As an alternative to solving the last polynomical equation of the 5th degree, we may transform the left member into a product by the command

```
syms x, factor(x^5-15*x^4+85*x^3-225*x^2+274*x-120)
```

which yields

```
(x-5)*(x-1)*(x-2)*(x-3)*(x-4)
```

making the five integral roots evident. If we replace the last term in the equation by 119, however, the *factor* command returns the same expression, indicating that it can't find exact factors.

The symbolic program permits us to manipulate algebraic expressions in several ways. For instance, if we type

```
syms x a, p=expand((x+a)* (x+2*a)* (x+3*a)* (x+4*a)* (x+5*a))
```

we obtain a result in the usual MATLAB style.

```
p=
  x^5+15*x^4*a+85*x^3*a^2+225*x^2*a^3+274*x*a^4+120*a^5
```

You could also have obtained this polynomial by long-hand, but only after some tedious work. In order to view it more clearly, you may add the command

```
pretty(p)
```

which yields the nearly-standard mathematical typesetting

```
 5       4           3  2          2  3          4       5
x  + 15 x  a + 85 x  a  + 225 x  a  + 274 x a  + 120 a
```

Let us now see if we can factor this complicated expression by typing the line

```
factor(p)
```

As if by magic, our original product re-appears. We may rest assured that the result was obtained by applying ordinary mathematical rules, but you could probably not perform the same operation by hand. There are even professional mathematicians who never have used the algorithm that is the key to this deed.

There are other commands that *simplify* complicated expressions. As an example, we may transform a rational expression such as

$$\frac{x^3 - 15x^2 + 71x - 105}{x^2 - 8x + 15}$$

by typing

```
simplify( (x^3-15*x^2+71*x-105)/( x^2-8*x+15) )
```

to obtain x-7.

Even transcendental expressions may be compacted, e.g.

```
simplify( sin(x)^2+ cos(x)^2 )
```

which yields the well-known trigonometric unity.

Using a more powerful command

```
simple( 64*cos(x)^7-112*cos(x)^5+56*cos(x)^3-7*cos(x) )
```

we obtain the final answer

```
cos(7*x)
```

which we can easily verify by *expand*. As you notice when applying *simple*, the program tries several transformations and finally presents the shortest answer, as you will see by pushing the *PgUp* button. The *help* command yields more details about these useful operations.

It frequently happens in mathematical work that one has to substitute a function for a variable in an expression. The symbolic mode of the program also offers a means of doing this, i.e.

```
subs( x*exp(3*x),  x, log(x) )
```

which replaces x in the first expression by log(x). The command *simple* and *pretty* help us reduce the answer to the short format

```
        3
log(x) x
```

Solving Systems of Linear Equations

The *solve* function also handles systems of equations. We may illustrate this by the simple example on p.30. It suffices to type the two equations with a separating comma, as follows.

```
syms x y, [x0 y0]=solve(x+2*y-8, 3*x+4*y-18)
```

This system has two unknowns, and hence we expect the *solve* command to yield two values as a vector [x y], in the style we have already seen on p.58. The solution is of course the same as what we obtained by the matrix formalism.

If we change the second constant from 8 to 8.1 and try to solve the system

```
syms x y, [x0 y0]=solve(x+2*y-8.1, 3*x+4*y-18)
```

The program then delivers the answers in the form of fractions of integers, since such values can't generally be expressed as decimal numbers.

```
x0 =
   9/5
y0 =
   63/20
```

The symbolic program yields exact numerical answers, but we could equally well introduce symbolic coefficients. For a general system of two linear equations we could write

```
syms x y a b c d e f, [x0 y0]=solve(a*x+b*y-c, d*x+e*y-f)
```

to obtain the solution

```
x0 =-(b*f-c*e)/(-b*d+e*a)
y0 = (-d*c+f*a)/(-b*d+e*a)
```

Solution by Matrices

Matrix operations are also available in the symbolic mode and are used in the same way as on p.31. For instance, to create a symbolic 2×2 matrix we could type

```
clear, syms a11 a12 a21 a22; A= [a11 a12; a21 a22]
```

and for the corresponding column vector

```
syms b1 b2; B= [b1; b2]
```

To solve this system of equations we may type, as before,

```
X0=A\B
```

which yields the solution vector

```
X0 =
   [ -(-a12*b2+b1*a22)/(-a11*a22+a21*a12)]
   [  (-a11*b2+a21*b1)/(-a11*a22+a21*a12)]
```

You may not necessarily obtain an expression of precisely the above form, but at least it is equivalent. To verify that the program gave us the correct answer we multiply the symbolic solution by A

```
left=A*X0, simplify(left)
```

which does return a column vector equal to B.

By the obvious command

```
Ai=A^-1
```

we find the inverse matrix Ai to be

```
[ -a22/(-a11*a22+a21*a12),  a12/(-a11*a22+a21*a12)]
[  a21/(-a11*a22+a21*a12), -a11/(-a11*a22+a21*a12)]
```

In addition to the above standard matrix mathematics, *element-by-element* operations remain valid. For instance, the expressions

```
A.*A, A.^A
```

yield the results we expect from the corresponding numeric examples (p.29).

New Function Plotter

For expressions with only one variable, the symbolic mode offers a convenient plot command, which we can use to inspect roots, extreme points, and so on. As a simple test, let us plot

$f(x) = \sin(x)^3 + \cos(3x)$

over one period by the "easy" command

```
clear, syms x, ezplot(sin(x)^3+ cos(3*x), [-pi pi], 1), grid on
```

Notice that we did not have to manipulate X and Y vectors, which makes this routine much simpler to use. The range of x may be given as a vector argument, but it is also possible to use the separate *axis* command. Finally, we need not specify figure(1) if we include the figure number as the last argument in parentheses. The expression automatically appears as the title above the plot, as seen below.

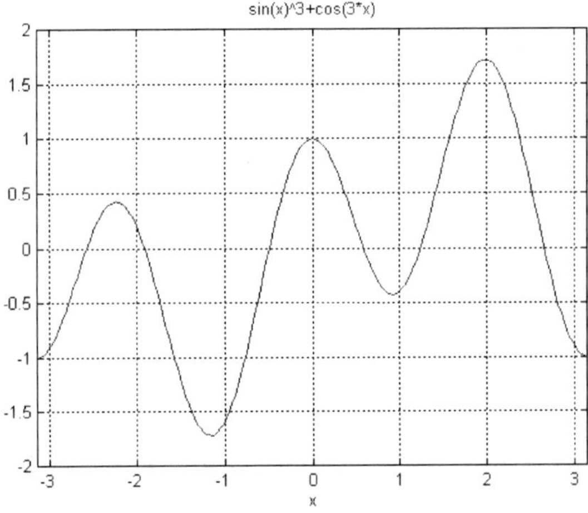

Numerical Approximation

Plotting is not really an example of symbolic algebra, since it involves numerical values presented on the screen. In fact, we may always convert an expressions having a definite value to a decimal number by a command known as *Variable Precision Arithmetic*, or *vpa*.

For example, we may write

```
clear, v=2*asin(1), v80=vpa(v, 80), vpa(pi,80)
```

where 80 is the number of digits demanded. We could then check the result by the line

```
vpa(sin(v80), 80)
```

Exercises

- ❑ Solve the equation $20x^2 - 31x + 12 = 0$ by the symbolic program.
- ❑ Solve $\cos(x) - x = 0$ by the symbolic program. How do you interpret the result?
- ❑ Try factoring the number 12341234.
- ❑ Expand the expression $(a + \sin(x))(b + \cos(x))(c - x)$ and then factor the result.
- ❑ Simplify $(x^{12} - y^{12})/(x - y)$ to obtain a polynomial. Then factor the latter expression.
- ❑ Solve the system with 3 unknowns on p.31. by the symbolic command.
- ❑ Use *ezplot* to display the function $y = \exp(-x^{50})$ over a suitable interval.
- ❑ Create a symbolic 3×3 matrix and find its inverse.

12 Limits

The function $f(x) = 1/x$ decreases steadily as x goes from just above zero toward plus infinity. Although it never reaches a final value, it comes as close to zero as you like as x increases. The mathematical way of expressing this fact is

$$\lim_{x\to\infty} \frac{1}{x} = 0.$$

Limits are extremely important concepts in calculus, which in turn is a universal tool in all the quantitative branches of science.

In this chapter we shall discover that we can find limits of many simple functions just by plotting them. It is true that MATLAB can't plot functions all the way to infinity, but we can make x as large as 10^{300}, which usually is more than sufficient to show the trend.

Limits Toward Large Values

In order to survey the variation of a function over a wide range we make use of a semi-logarithmic plot, where successive powers of ten are equally spaced on the horizontal (x) axis, as in the next figure.

As the first example we inspect the expression

$$\lim_{x\to\infty} \frac{2x-1}{\sqrt{3x^2+x+1}}$$

by the file below.

```
% ex121.m: Limit Toward Infinity
clear,  echo off
X=logspace(0, 100, 600);            % Range of exponents, 600 points
Y=(2*X-1) ./ sqrt(3*X.^2+X+1);
figure(1),  semilogx(X,Y),  grid on,  zoom on,  xlabel('x')
```

The new command *logspace* creates a vector X with 600 values from 10^0 to 10^{100}, and *semilogx* puts the corresponding scale on the horizontal axis of the plot, as shown below.

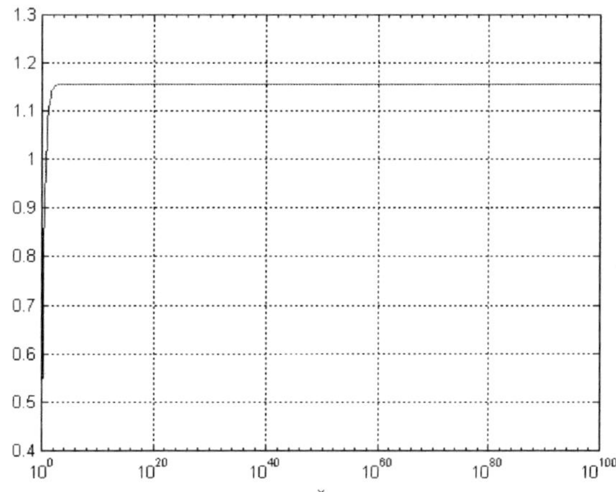

We could have extended the above curve for values of x up to 10^{300}, but then it would have been difficult to discern the rise at the left end. The curve seems to be completely flat over most of the range, and we may study the limiting behavior further by clicking on the right end of the curve to zoom it. To illustrate the constancy by numeric values we could also type

```
format long,  Y(300),  Y(600)
```

after the run to read the middle and last values of the vector. If we divide both numerator and denominator by x, we easily see that the limit must be $2/\sqrt{3}$. You may verify that against our numerical results.

Of course, we can also observe the behavior of an expression toward *minus* infinity. For instance, let us study the expression

$$\lim_{x\to-\infty} \frac{\sqrt{1-x}\left(1-\sqrt{3-2x}\right)}{1+6x}$$

by the file

```
% ex122.m:  Limit Toward Minus Infinity
clear,  echo off
X= -logspace(1, 300, 600);                         % Make 600 negative values
Y= (sqrt(1-X).*(1-sqrt(3-2*X)))./(1+6*X);
figure(1),  semilogx(X,Y),  grid on,  zoom,  xlabel('x')
```

The following plot results from running this file. Again we may verify the limiting value, and by a similar analysis.

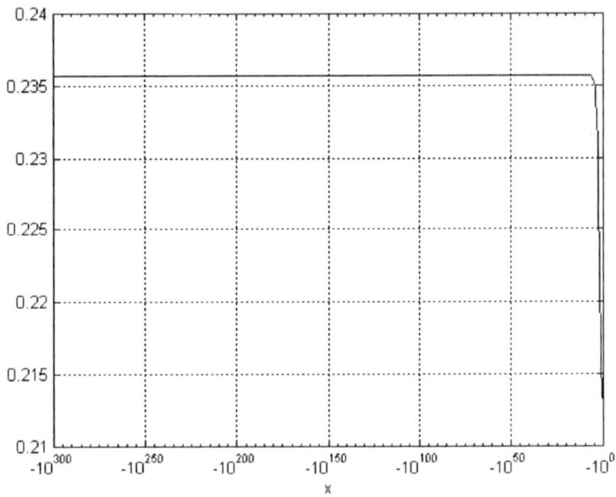

As a final example of limits toward infinity we take

$$\lim_{x\to\infty} \frac{x}{\ln(x^2)}$$

or in MATLAB notation

```
% ex123.m:  Limit Toward Plus Infinity
clear,  echo off
X= logspace(1, 100, 600);  Y=X./log(X.^2);
figure(1),  semilogx(X,Y),  grid on,  zoom,  xlabel('x')
```

Here we obtain a curve that climbs violently near the right end, indicating that the limit is $+\infty$.

Limits Toward Zero

One might think that a linear x-axis would yield useful plots in this case, but as you find when trying a linear plot, this is not very informative. Thus we adhere to the logarithmic scale but use *negative* powers. We study the classic example

$$\lim_{x\to 0+} \frac{\sin(x)}{x}$$

where we approach zero from the positive side. The script file for this case could read

```
% ex124.m: Limit Toward Zero
clear, echo off
X= logspace(-50, 0, 600); Y= sin(X)./ X;
figure(1), semilogx(X,Y), grid on, zoom, xlabel('x')
```

The plot below shows that the expression tends to 1 as x goes toward zero.

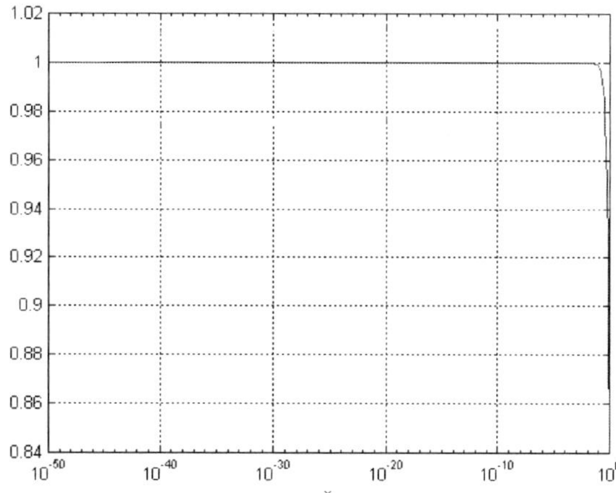

In summary, a direct plot illustrates the fundamental idea of a limit, and if a finite limit exists we can usually calculate it numerically.

Exercises

❑ Study the case of $\lim_{x \to 0+} \sin(x)/x$ by a linear plot over the interval $0 < x < 1$.

Make plots to find the following limits.

❑ $\lim_{x \to +\infty} \sin(x)/x$ for $10^0 < x < 10^5$

❑ $\lim_{x \to +\infty} x \sin\left(\frac{1}{x}\right)$ for $10^0 < x < 10^{10}$

❑ $\lim_{x \to +\infty} \exp(1/x)$ for $10^0 < x < 10^{10}$

❑ $\lim_{x \to +\infty} \left(1 + \sin\frac{3}{x}\right)^x$ for $10^0 < x < 10^{10}$

❑ $\lim_{x \to 0+} x^{\sqrt{x}}$ for $10^{-50} < x < 10^0$

13 Derivatives

At any point of one of the curves we plotted earlier we can imagine drawing a tangent line. The derivative could be defined as the slope, $\Delta y/\Delta x$, of that line at the point chosen. The figure on the next page illustrates this concept.

First-Order Derivative

The derivative is formally defined by

$$f'(x) \equiv \lim_{h\to 0} \frac{f(x+h)-f(x)}{h} \equiv \lim_{h\to 0} q \qquad \bullet$$

Let us take the function $\sin(x)$ as an example of how to calculate the derivative $f'(x)$ numerically.

```
% ex131: Derivative as a Limiting Quotient
clear, echo off
X= 1.0:1e-3:1.5; Y= sin(X);
figure(1), plot(X,Y), grid on, xlabel('x'), ylabel('sin(x)')
x=1.2;                                   % Value of x for the derivative
H=logspace(-10, 0, 500);                 % Vector of values h
Q=(sin(x+H)- sin(x))./H;                 % Vector of quotients q
figure(2), semilogx(H,Q), grid on, xlabel('h'), ylabel('q')
format long, limiting_q=Q(1), derivative=cos(x)      % Comparison
```

The figure below is an edited version of the first plot. It indicates the point x where we shall calculate the derivative and includes the corresponding tangent. It also shows the increment Δf caused by the change h in the independent variable x.

We now wish to plot the quotient q as a function of h, as the latter tends to zero. For this purpose, we store the values of h in the vector H (ranging from 10^{-10} to 10^{0}) and the corresponding values of the quotient q in the vector Q. This permits us to plot as we did in the preceding chapter for finding a limit.

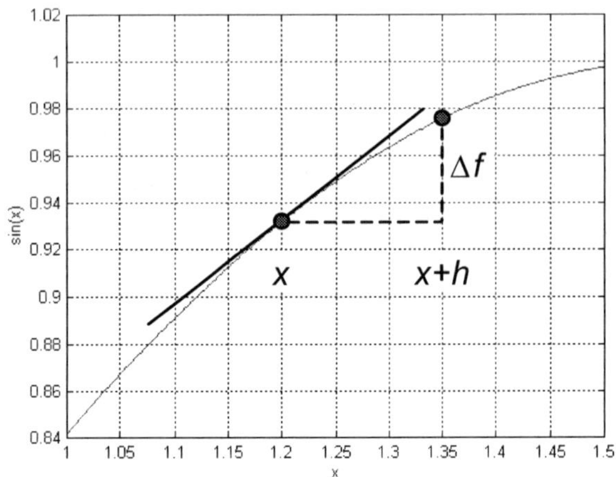

Running the above script file we obtain the next plot. The quotient seems to reach a limit at about 0.36 for the smallest values of h. On the last line of the script we compare this limit to the value of the exact derivative, and we do find agreement in the first six figures.

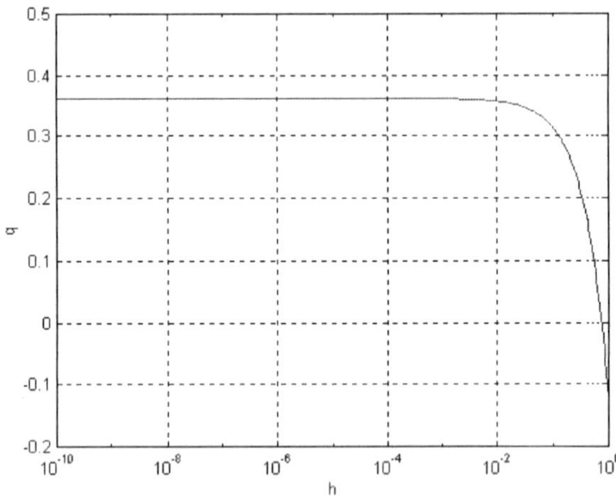

We have thus obtained the derivative at one point, but we are free to introduce other values of x. It would be rather time-consuming, however, to plot the derivative of a function by the above method. In the following, we shall thus choose a *fixed* value of *h*, and verify by comparison that it was sufficiently small.

We are now ready to plot first derivatives of functions, which is almost as easy as plotting the functions themselves. Let us start by writing a function file as follows.

```
% xcos.m:  Function for Derivatives
function F= xcos(X);
F=  X.*cos(X);
```

The script file below includes the derivative quotient according to the definition. MATLAB allows us to add the scalar *h* directly to the vector X, thereby adding *h* to all of its elements. We have chosen the value 1e-4 for the increment, just on the basis of experience from the preceding section. We shall soon see if it is small enough.

```
% ex132.m:  Plot a Function and its Derivative
clear,  echo off
X=0: 0.01: 10;  Y=xcos(X);
h= 1e-4;                                   % Fixed increment
Der=(xcos(X+h)- xcos(X))/ h;          % Quotient
figure(1),  plot(X,Y,  X,Der,'--'),  grid on,
   legend('f=x*cos(x)', 'df/dx'),  xlabel('x')
Der_ex=cos(X)-X.*sin(X);                   % Exact derivative
figure(2),  plot(X,Der,  X,Der_ex),  grid on,  legend('df/dx', 'exact')
figure(3),  plot(X, Der- Der_ex),  grid on,  title('Error in df/dx')
```

In this script file we have requested three plots, which may be selected by the Windows task buttons and viewed. In the first one (below) we have dashed the curved corresponding to the derivative, to make it recognizable in black and white. The character string at the end of the plot parentheses provides this modification of the curve. The command *help plot* lets us know about other alternatives. We have also dragged the *legend* to the left by the mouse.

The second plot (not shown) compares the numeric derivative to the exact one. The two curves lie too close to be distinguished. Since the second curve overwrites the first one, the color becomes that assigned to Der_ex.

Because of the overlapping in the second plot we include the third one (not shown), which displays the error with respect to the exact expression. This error turns out to have a maximum of about 5e-4.

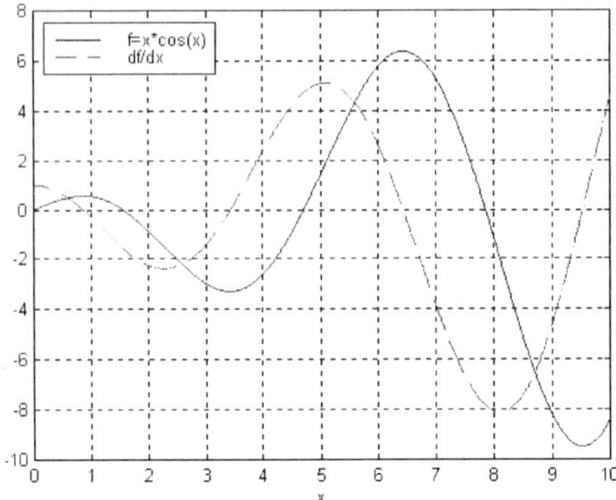

So far we have used positive values of h when calculating a derivative, which means that we obtain the limit as $h \to 0+$. We could equally well have used a negative value, however, to obtain the derivative to the left. At large values of h these two approximations to the derivative may, however, be noticeably different.

It is usually better to use an *average* of the left and right approximations to the derivative, which may be written

$$f'(x) \equiv \lim_{h \to 0} \frac{f(x+h) - f(x-h)}{2h} \qquad \bullet$$

This could be called a *symmetric* approximation for the derivative.

To test this alternative formula we need only change the name of the script file and replace the line containing the quotient, as shown below.

```
% ex133.m:  Plot a Function and its Symmetric Derivative
...
Der=(xcos(X+h)- xcos(X-h))/(2*h);                % Symmetric quotient
...
```

The next figure is the resulting plot of the error in the derivative. This error is now of the order of 1e-8, although the increment h is unchanged. We conclude that the symmetric formula is to be preferred.

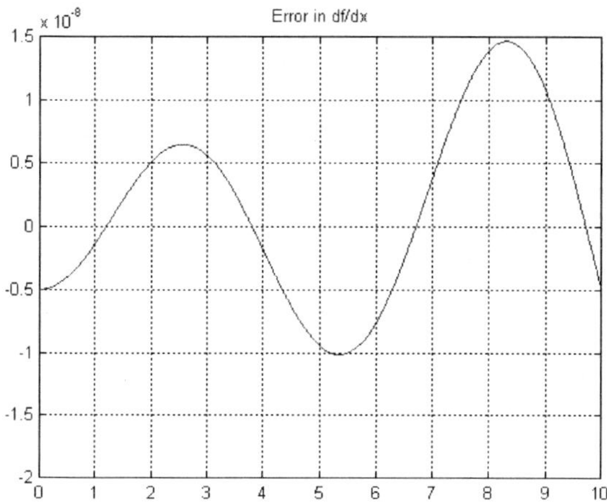

Second-Order Derivative

When analyzing the behavior of a function it is of some interest to know not only the first but also the second derivative. The latter is the derivative of the derivative, and using the symmetric recipe it may be written

$$f''(x) \equiv \lim_{h \to 0} \frac{f'(x+h) - f'(x-h)}{2h} =$$

$$\lim_{h \to 0} \frac{1}{2h} \left[\frac{f(x+2h) - f(x)}{2h} - \frac{f(x) - f(x-2h)}{2h} \right] =$$

$$\lim_{h \to 0} \frac{f(x+2h) - 2f(x) + f(x-2h)}{4h^2},$$

or if we change notation by substituting $2h \to h$

$$f''(x) = \lim_{h \to 0} \frac{f(x+h) - 2f(x) + f(x-h)}{h^2}$$ ●

We may incorporate this expression into the preceding script file, and also compare the result to the exact expression. Using *ex132* as a template, we modify it as follows.

```
% ex134.m:  Plot Second-Order Derivative
clear,  echo off,  close all                        % Close earlier figures
X=0:0.01:10;  Y=xcos(X);
h=1e-4;                                             % Fixed increment
Der=(xcos(X+h)- xcos(X-h))/(2*h);                   % Symmetric quotient
Der2=(xcos(X+h)-2*xcos(X)+xcos(X-h))/h^2;                 % d2f/dx2
figure(1),  plot(X,Y,  X,Der,'--',  X,Der2,':'),  grid on,  xlabel('x')
  legend('f=x*cos(x)', 'df/dx', 'd2f/dx2')
Der2_ex=-sin(X)- sin(X)- X.*cos(X);                 % Exact d2f/dx2
figure(2),  plot(X,Der2,  X,Der2_ex),  grid on
  legend('d2f/dx2', 'exact')
figure(3),  plot(X,Der2-Der2_ex),  grid on,  title('Error in d2f/dx2')
```

It may happen that a feature of an earlier figure, such as a legend, may remain on the screen. For this reason we issue the command *close all* to remove all remaining graphics.

After executing this file we may view the first figure comprising the function and its 1st and 2nd derivatives, both computed by symmetric algorithms. We obtain a dotted curve for the 2nd derivative by including a string after the plot coordinates.

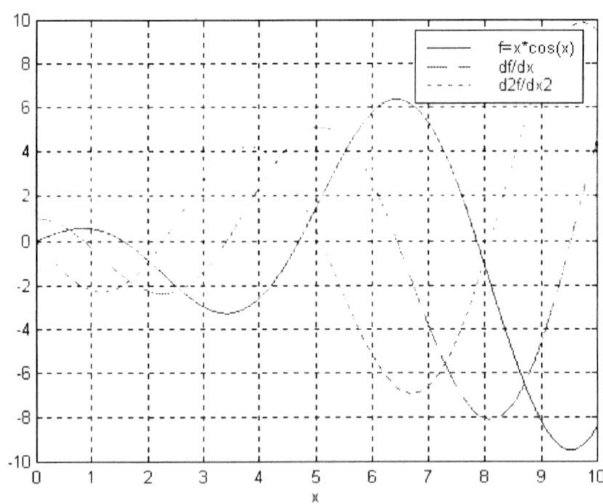

The third figure, shown below, indicates that the error in the numerical 2nd derivative is as low as a few times 1e-7. Here, we encounter a new type of deviation, which has a random component, caused by numerical round-off errors. This scatter vanishes if we put $h = 1e - 3$, but then the error in the 1st derivative increases.

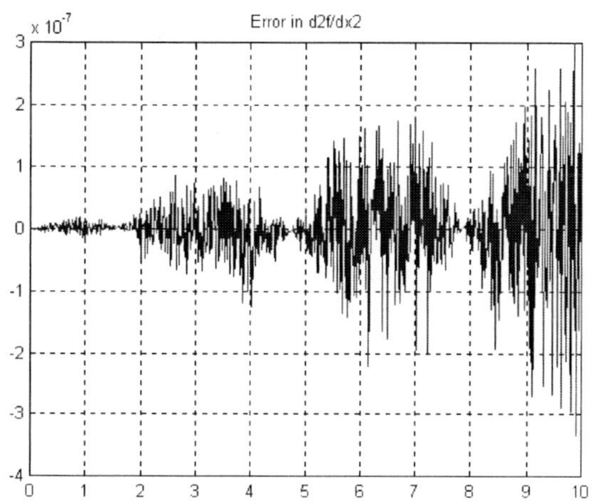

Exercises

❑ Returning to *exp132*, how much is the error in the derivative reduced if you decrease h by a factor of 10 in the simplest approximation?

❑ How much does the error change if you first increase, then decrease h in the *symmetric* quotient for by a factor of 10?

❑ Study the derivative of the Gaussian function, $\exp(-x^2)$, after the model of *ex132*. Create a new function file and choose a suitable range of x for the plots.

14 Sums

It often occurs in applied mathematics that one wishes to sum over terms depending on an index. Such repetitive operations are ideal tasks for a computer. A familiar example is

$$S(n) = \sum_{i=1}^{n} k^{i-1} \equiv k^0 + k + k^2 + k^3 + \ldots + k^{n-1} = \frac{1-k^n}{1-k} \qquad \bullet$$

which is known as a geometric series. The final expression is the analytic sum of this series. Let us now calculate it numerically for a given quotient k.

The core of the following script file is the manipulation of the current sum si. The *for* construct employed is somewhat similar to *while*, but here we step the index over a fixed range.

We first set si to zero and then add the various terms successively. It is the statement si=si+k^(i-1) that carries out this accumulation of terms. Taken as a conventional mathematical equation, si=si+k^(i-1) is of course nonsense, since it would imply that k^(i-1) is zero. We must think of it as changing the *contents* of the variable si into si+k^(i-1) by adding a term.

```
% ex141.m:  Sum of a Geometric Series
clear,  echo off,  format long,
k=input('k= ');                            % Enter k from keyboard
si=0;                                      % Initial value for sum
for i=1:100                                % Set up loop
  term=k^(i-1);                            % Term with index i
  si=si+term;                              % Add term to si
  si_ex=(1-k^i)/(1-k);                     % Analytic expression for si
  sums=[si  si_ex],  pause                 % Display, continue by return
end                                        % End of loop
```

If we enter k=2 and step through successive values of the index i, we find that the numerical sum matches the analytic one perfectly. Increasing to k=10, we find similar agreement but with larger numbers.

The file also works for values of k smaller than unity. For instance, if we take k=0.5 we find that the sum approaches a limiting value of 2.0.

Convergent Series

It is obvious from the analytic expression for the sum of the geometric series that

$$\lim_{n\to\infty} \frac{1-k^n}{1-k} = \frac{1}{1-k}$$ ●

if $k < 1$. This series thus tends to a limit and is known as a *convergent* series. The difference between this type of limit and that discussed in Chapter 12 is that the variable n takes only integral values.

We shall now modify *ex141* to *plot* the sum as a function of the index. In order to prepare for large indices in some cases we use the *while...end* loop again, as on p.63. Here, we must arrange for the index i to step to the next integer for every turn, which we do by the statement i=i+1. This is similar to summing the terms, but here the term is unity. As with error on p.64, we now break the loop when term has become sufficiently small.

```
% ex142.m:  Sum of a Geometric Series
clear,  echo off,  format long,
k=input('k= ');                         % Enter k from keyboard
si=0;  i=0;                             % Initial values for sum and index
while 1==1                              % Set up loop
  i=i+1;                                % Next index
  term=k^(i-1);  si=si+term;
  Index(i)=i;                           % Index=vector of indices
  Sum(i)=si;                            % Sum=vector of sums
  if abs(term)<1e-15, break,  end
end                                     % End of loop
figure(1),  plot(Index, Sum),  grid on,  xlabel('index'),  ylabel('sum')
  title('Geometric Series')
sum_ex=1/(1-k);                         % Analytic limit
sums=[Sum(i) sum_ex]                    % Last sum compared
```

Running this file with k equal to 0.5, then 0.1 and 0.9, we obtain a set of rather different results, but the sum invariably converges to the expected limit. The plot below is for k=0.9.

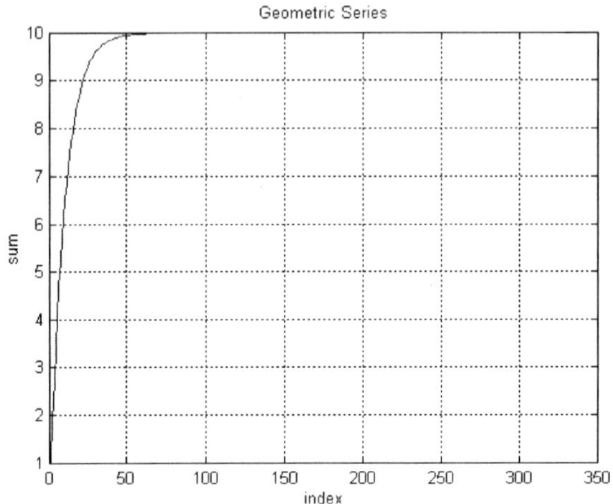

Divergent Series

Our next example is completely different, and no closed expression for the sum is available. We shall try summing

$$S(n) = \sum_{i=1}^{n} \frac{1}{i}$$

using a script file similar to *ex142*. Here, we revert to the for...end loop to sum over a given number of terms.

```
% ex143.m:  Sum of 1/i
clear,  echo off,  format long
si=0;
for i=1:1000                              % Set up loop
  term= 1/i;  si=si+term;                 % Add the term
  Index(i)=i;                             % Index=vector of indices
  Sum(i)=si;                              % Sum=vector of sums
end                                       % End of loop
figure(1),  plot(Index, Sum),  grid on,  xlabel('index'),  ylabel('sum')
  title('Sum of 1/i')
```

The resulting plot does not exhibit the flattening typical of convergence, and it even looks much the same if we let the index run up to 5000. Hence, we assume the present series to be divergent.

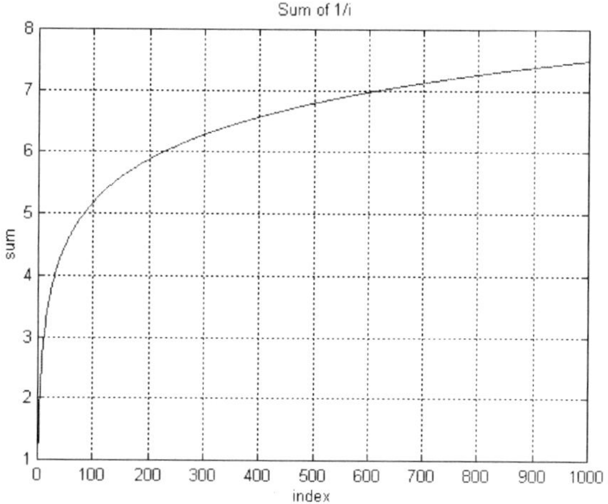

This curve may remind you of one you have already seen in this book. In one of the exercises you are invited to explore if that is correct.

Trigonometric Series

As any textbook on calculus will tell you, the sine and cosine functions may be expressed as infinite power series. The one for $\sin(x)$ is

$$\sin(x) = \sum_{i=0}^{\infty} (-1)^i \frac{x^{2i+1}}{(2i+1)!} \equiv \frac{x}{1!} - \frac{x^3}{3!} + \frac{x^5}{5!} - \ldots \qquad \bullet$$

where $n! \equiv n(n-1)(n-2)\ldots 3 \times 2 \times 1$, known as the *factorial* of *n*. When summing this series we may of course calculate the factorial for each term, but there is a simpler way. We may calculate the first term and then multiply by the appropriate factor to obtain the second one, and so on. The ratio t_i / t_{i-1} of two successive terms is found to be

$$\frac{t_i}{t_{i-1}} = \frac{(-1)^i x^{2i+1}}{(2i+1)!} \frac{(2i-1)!}{(-1)^{i-1} x^{2i-1}} = \frac{-x^2}{(2i+1)2i}$$

Using *ex142* as a template we type the following file. Since we shall need to express the error, which is extremely small, we demand scientific notation.

```
% ex144.m:  Power Series for sin(x)
clear,  echo off,  format long;
x=input('x= ');
term=x;  s=x;  i=0;                          % Initial values
while 1==1                                   % Forever
  i=i+1;                                     % Next index
  term=term*(-x^2)/(2*i+1)/(2*i);
  s=s+ term;
  if abs(term)<1e-15;  break;  end;
end
last_index_in_sum=i
sum=s,  sin_ex=sin(x)
format short e,  error=s- sin(x)             % Scientific notation
```

This evaluation evidently requires surprisingly few terms, no matter what value of x we choose in the interval $-\pi < x < \pi$, and the sum is in fact correct to the last of the 14 digits displayed. It thus seems easy to approximate the value of an infinite power series that is known to be convergent.

Exercises

❑ Estimate the value of the sum $\sum_1^\infty i^{-4}$ and compare to $\pi^4/90$.
❑ The same for the terms $1/i/(i+1)$ from i=1 to infinity. Compare to 1.0.
❑ Compute the sum of $(-1)^i/(2i+1)$ from i=0 to infinity. Compare to $\pi/4$.
❑ Modify *ex143* by adding a plot of the difference si-log(i).

15 Integrals

The integral of a function over an interval of x is simply the area between the corresponding curve and the x-axis. The formal definition by Riemann regards it as the limit of the area for a flight of stairs approximating the curve. The stairs connects to the curve as shown in the figure, and the limit is taken for very small step-length.

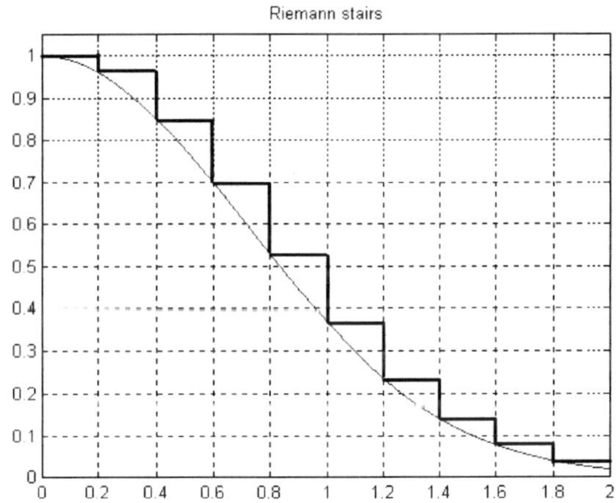

To replace the curve by a series of stairs obviously means that we sum the areas of a number of rectangles. In this particular case, the area under the stairs evidently is larger than the area under the curve, but that would not be true for any functional shape.

Stairs Integration

We shall now approximate the integral by the formula

$$\int_{x_1}^{x_n} f(x)dx \cong \sum_{i=1}^{n-1} f(x_i)\,h \equiv \{f(x_1)+f(x_2)+f(x_3)+\ldots+f(x_{n-1})\}\,h \qquad \bullet$$

where n is the number of function values and h the step-length. If we express the crucial values in curly brackets as a vector, we may exploit the automatic summing provided by MATLAB. As an example we choose to integrate $\cos(x)$ over an interval starting at $x = 0$. This simplest strategy of integration only requires the following script lines.

```
% ex151.m:  Integration by Direct Summing
clear,  echo off,  format long
xn=input('xn= ');                    % Right end coordinate
x1=0;  n=500;                        % Number of sub-intervals
h=(xn-x1)/n;                         % Length of sub-interval
X=x1:h:xn-h;                         % Omit the last point
F=cos(X);
integral=sum(F)* h;                  % Sum of function values
int_ex=sin(xn)-sin(x1);              % Exact integral
integrals=[integral  int_ex]         % Display comparison
format short e,  error=integral-int_ex
```

Here, we may choose xn freely from the keyboard. For instance, we may test a few values such as 1, pi/2, 2, 3, pi, 4, 5.

The result was not much to write home about, was it? Since the sum converges to a limit, however, you may improve things by using n=5000 or even 16000, which is close to the limit allowed by the Student Edition.

Trapezoidal Integration

Honestly, the stairs approximation to a continuous curve is not the best one can imagine. Bridging each interval by a straight line from one point to the next should be more accurate. Instead of summing areas of rectangles we then effectively divide into trapezoids as follows.

$$S = \frac{f(x_1) + f(x_2)}{2}h + \frac{f(x_2) + f(x_3)}{2}h + \ldots + \frac{f(x_{n-1}) + f(x_n)}{2}h$$

Adding the two first terms we realize that the result may also be written

$$S = \{f(x_1)/2 + f(x_2) + f(x_3) + \ldots + f(x_n)/2\}h \qquad \bullet$$

The difference with respect to the Riemann sum is that all of the n function values are used, but the first and last values are divided by two. In other words, we may calculate the integral as before (with the last term included) and then subtract the expression

$$\frac{f(x_1)+f(x_n)}{2}h$$

from the sum. The new script file will thus look like this.

```
% ex152.m:  Trapezoidal Integration
clear,  echo off,  format long
xn=input('xn= ');                    % Right end coordinate
x1=0;  n=500;                        % Number of sub-intervals
h=(xn-x1)/n;                         % Length of sub-interval
X=x1:h:xn;  F=cos(X);
integral=sum(F)* h - (F(1)+ F(n)) /2*h;
int_ex=sin(xn)-sin(x1);              % Exact integral
integrals=[integral  int_ex]         % Display comparison
format short e,  error=integral-int_ex
```

You will find that the results now are much improved. Using this file as a template you could also integrate another function by defining it in a separate file, replacing cos in the script file by the new name.

Quadratic Approximation

Trapezoidal approximation is an excellent method for integrating functions that should be continuous but do not have a derivative, such as data measured at discrete points. For mathematical functions that may be differentiated an unlimited number of times, such as $\cos(x)$, more efficient algorithms are available. We shall now try the one due to Simpson.

The essence of Simpson's method is to approximate the function by second-degree polynomials (parabolas). As before, we subdivide the integration interval into equal parts. To the first three points, x_1, x_2 and x_3 say, we fit a second-degree polynomial, and similarly to x_3, x_4 and x_5, etc.. Every parabola segment thus bridges two intervals, so the total number of intervals has to be even.

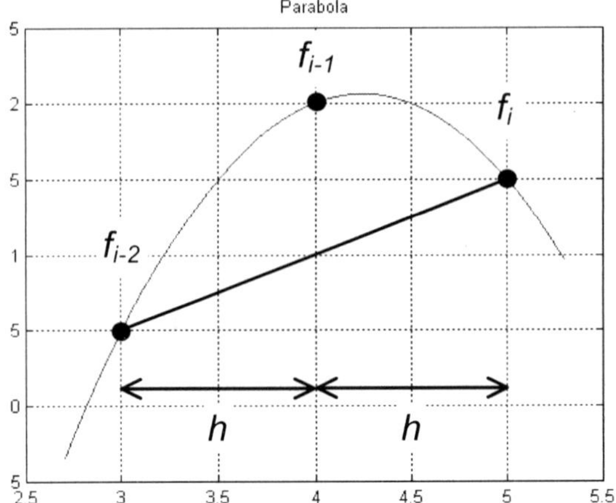

As we see from the above figure, the area under the curve from x_{i-2} to x_i may be divided into two parts, a trapezoidal area below and a parabolic area above. The trapezoidal part of the area obviously is $(2h)(f_{i-2}+f_i)/2$.

The parabola must be symmetric with respect to x_{i-1}, since it passes through zero at both ends of the interval. Its height at the apex is $f_0 = f_{i-1} - (f_{i-2}+f_i)/2$. You may show by elementary integration that the area under the concave parabola is $4f_0h/3$. The sum of the two areas spanning the twin interval will thus be equal to

$$(f_{i-2} + 4f_{i-1} + f_i)h/3, \qquad \bullet$$

where h is the length of a single interval.

We shall use this formula successively to accumulate the integral from x_1 to any odd-numbered point, such as x_3, x_5, x_7, etc.. In the following script file we add the areas of successive twin intervals and store the intermediate results in the variable sum.

```
% ex153.m:  Simpson Integration
clear,  echo off,  format long
xn=input('xn= ');                    % Input right end coordinate
x1=0;  n=500;                        % Even number of sub-intervals
h=(xn-x1)/n;                         % Length of sub-intervals
X=x1:h:xn;                           % Interval coordinates (n+1)
sum=0;                               % Reset sum
for i=3:2:n+1                        % For i=3,5,7,...n+1
  sum=sum+ cos(X(i-2))+ 4*cos(X(i-1))+ cos(X(i));        % Add area
```

```
end
integral=sum*h/3;
int_ex=sin(xn)-sin(x1);          % Exact integral
sums=[integral int_ex]           % Display comparison
format short e,  error=integral-int_ex
```

When you test this file with various values of xn, you will probably be pleasantly surprised at the small error.

Improper Integrals

So far we have integrated a function over a finite interval of x. In many cases the integral values tend to a limit as x_n increases toward plus infinity, or x_1 toward minus infinity. We can treat such integrals, known as improper, in the same way as did with sums (p.89). To be specific, we regard the Simpson integral over a twin interval as if it were a term in a sum.

In the following script file we replace the cos function by gaussian, which we defined on p.56. Here, we consider one Simpson interval $(x_1 ... x_3)$ and shift it to higher x-values on going to the next turn of the loop.

The logic of this file differs a little from that of *ex153*. Here, we fix the interval length and increment x1 at the end of the loop to obtain the new interval coordinates.

```
% ex154.m:  Improper Integral
clear,  echo off,  format long
h=0.01;                          % Length of sub-intervals
x1=0;                            % First Simpson coordinate
sum=0;  i=0;                     % Reset sum and index
while 1==1                       % Forever
  i=i+1;                         % Step up index
  x2=x1+h;  x3=x1+2*h;           % Simpson coordinates
  term=gaussian(x1,1)+4*gaussian(x2,1)+gaussian(x3,1);
  sum=sum+term;                  % Sum of areas
  X(i)=x3;                       % Coordinate for plot of integral
  Int(i)=sum*h/3;                % Store integral as vector element
  if abs(term/sum)<1e-14,  break,  end
  x1=x1+2*h;                     % Prepare x1 for next turn
end                              % End of loop
```

```
figure(1), plot(X,Int), grid on, xlabel('x'), ylabel('integral')
i                                  % Number of loops used
integrals=[Int(i) sqrt(pi)/2]      % Comparison to exact
format short e, error=Int(i)-sqrt(pi)/2
```

Much as we did with series we now stop the calculations when the term has become sufficiently small compared to the sum.

The curve (below) displayed by this file clearly indicates that the integral converges to a limiting value, and the numeric comparison shows that the latter is extremely close to the exact expression, $\sqrt{\pi}/2$.

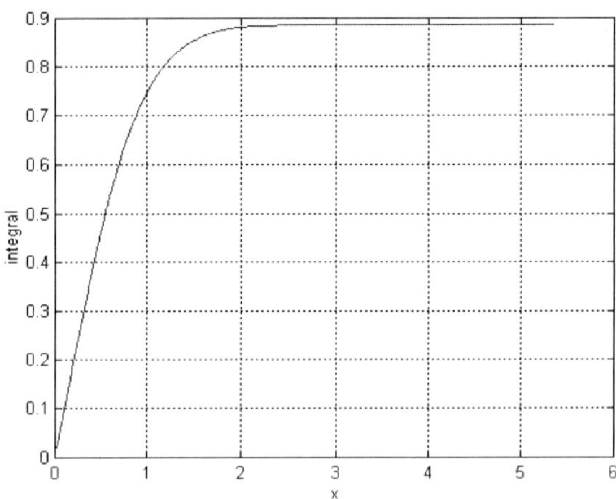

Exercises

❑ In the stairs approximation we never used the last function value, corresponding to x_n. We could just as well have omitted the first point instead. Find the formula for the average of these two alternative sums. Have you seen that expression before?

❑ Calculate the following integrals by Simpson approximation. Compare to results obtained with a smaller value of h.

$$\int_0^{\pi} \sqrt{\sin(x)}\, dx \qquad \int_0^{\pi} \sin\left(\sqrt{x}\right) dx \qquad \int_{-10}^{0} \frac{1}{1-x} dx$$

❑ Estimate the following integrals by Simpson's method. Compare to results obtained with a smaller value of h. Also try a few different stop conditions.

$$\int_0^{+\infty} x\exp(-x^2)\,dx \qquad \int_1^{+\infty} \frac{\ln(x)}{x^3}\,dx$$

16 Symbolic Calculus

There are general rules for finding the derivative of a function. Anyone who can learn to expand polynomial factors could also apply the procedures of differentiation by longhand. Just as in the case of polynomials, however, the number of operations is often staggering. Fortunately, any task involving tedious routine calculus may safely be left to a computer program.

Limits

The symbolic part of MATLAB readily applies the rules of limits. For example, let us revisit the expression from p.76

$$\lim_{x \to \infty} \frac{2x-1}{\sqrt{3x^2+x+1}}$$

To find the analytic answer we type

```
clear,  syms x,  limit( (2*x-1)/sqrt(3*x^2+x+1), x, inf, 'left')
```

The last argument ('left') may seem absurd, since we are requesting the limit to *plus* infinity, but x should in fact approach infinity from the left side. The result is what we could expect from the plot.

For the expression

```
limit( x/log(x^2), x, inf, 'left')
```

the answer is inf, as the plot again leads us to suspect. Here, a positive sign is understood. The line

```
limit( (1-x)/log(x^2), x, inf, 'left')
```

on the other hand, generates the answer -inf.

In a similar way, we easily find the limit of $\sin(x)/x$ for $x \to 0+$ by the command line

```
limit( sin(x)/x, x, 0, 'right')
```

The result is equal to 1, and we obtain an identical answer using left.

Derivatives

To obtain an expression for a derivative with respect to x we only need to type

```
clear, syms x, diff( exp(x*sin(x)), x)
```

The result is virtually instantaneous.

```
(sin(x)+x*cos(x))*exp(x*sin(x))
```

The program also produces the second derivative elegantly by

```
diff( exp(x*sin(x)), 2); pretty(ans)
```

```
                                                          2
(2 cos(x)  - x sin(x)) exp(x sin(x)) + (sin(x) + x cos(x))  exp(x sin(x))
```

When the function contains more than one independent variable, the program assumes that we mean to differentiate with respect to x, y or z, rather than a, p or q. This rule makes it simpler for us in a case such as

```
syms x a, f= x^a*sin(1/x); fx= diff(f)
```

but if we really mean to differentiate with respect to the variable a, we must type explicitly

```
diff(f,a)
```

which yields a completely different result.

Taylor Expansions

Any function possessing a sufficient number of derivatives may be expanded in a Taylor series, in terms of ascending derivatives at a point of reference. The symbolic program assumes x=0 as the default reference point. To generate such a series could not be more convenient. With the function $g = \exp(x\sin(x))$ we type the line

```
clear, syms x, g=exp( x*sin(x)); t= taylor(g,12)
```

to obtain the expansion until the 12th power.

```
t =
   1+x^2+1/3*x^4+1/120*x^6-11/560*x^8-1079/362880*x^10
```

We need only form the 10th derivative of g to realize how much drudgery we have avoided.

A question that conventional mathematics only partly answers, is how large the difference is between a function and its Taylor approximation. We are now in a position to assess the accuracy of the expansion by direct comparison in a plot.

```
% ex161.m: Compare Taylor Expansion with Function
clear, echo off
syms x, f=exp(x*sin(x));
t= 1+1*x^2+1/3*x^4+1/120*x^6-11/560*x^8-1079/362880*x^10;
ezplot(t, [0 2]), grid on, hold on, pause    % Continue by return
ezplot(f, [0 2]), grid on, hold off
```

Here, we superimpose two figures in the same diagram by the save command *hold on*. This file produces the following figure, which shows that the Taylor polynomial diverges rapidly (downwards) from the function for $x > 1.5$.

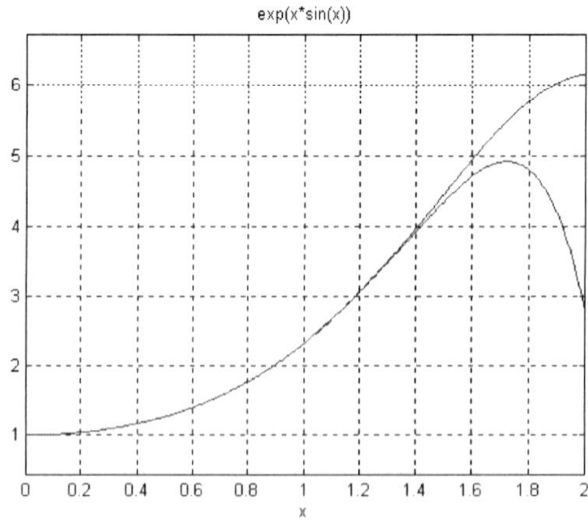

This behavior of the Taylor expansion is rather typical, in the sense that it follows the function very closely over a certain interval and diverges rapidly outside.

The above is really a special case of the Taylor expansion. We may use the same command to expand around a value different from zero, e.g. $x = 1$. This we accomplish by the line

```
g=exp( x*sin(x)); t=taylor(g,12, x, 1)
```

resulting in a horribly complicated answer. We may approximate the constants involved, however, to see more clearly what this series means. Let us exploit *Variable Precision Arithmetic* (p.74) by typing

```
vpa(t,3)
```

to obtain the coefficients to 3 digits. The result is

```
ans =
-.88+3.20*x+2.48*(x-1.)^2+.217*(x-1.)^3-1.13*(x-1.)^4-1.12*(x-1.)^5-
.336*(x-1.)^6+.219*(x-1.)^7+.281*(x-1.)^8+.106*(x-1.)^9-.295e-1*(x-
1.)^10-.522e-1*(x-1.)^11
```

You could now plot the *full* expansion about $x = 1$ for comparison.

Sums

The symbolic program also allows us to sum over terms depending on an index, for instance

$$\sum_{m=1}^{n} k^{m-1}$$

The use of i as the index of summation must be avoided, because the symbolic variable i has a special significance (imaginary unit). We obtain an expression for this well-known sum by

```
clear, syms k m n, symsum( k^(m-1), m, 1, n); simplify(ans)
```

which means that the index m is to run from 1 to n. The prompt answer

```
(k^n-1)/(k-1)
```

is what we expect.

We may even sum a convergent series to infinity, e.g.

$$\sum_{m=1}^{\infty} \frac{1}{m^4}$$

by the command line

```
syms m,  symsum( m^-4, m, 1, inf);  simplify(ans)
```

to obtain

```
1/90*pi^4
```

which we used for comparison in an exercise on p.92.

Indefinite Integrals

The symbolic program also helps us to find the indefinite integral (primitive function or anti-derivative), denoted

$$F(x) = \int f(x)dx$$

and defined by

$$\frac{dF(x)}{dx} = f(x)$$

To find the indefinite integral

$$F(x) = \int x^3 \cos(x)\, dx$$

we just type

```
clear,  syms x,  f= x^3*cos(x);  F= int(f);  pretty(F)
```

The result

```
 3              2
x   sin(x) + 3 x   cos(x)  - 6 cos(x)  - 6 x sin(x)
```

does not include a constant of integration, so we shall always have to imagine a constant C being added to the answer.

As a second example, let us take the integral

$$\int \frac{dx}{x(x^2 - a^2)}$$

which contains a parameter a. The command line

```
clear, syms x a, f=1/(x*(x^2-a^2)); F=int( f); Fs=simple(F), pretty(Fs)
```

yields the result

```
     -2 log(x) + log(a - x) + log(a + x)
1/2 -----------------------------------
                      2
                     a
```

Here the program assumes that we want to integrate with respect to x, rather than a, but we could also have declared the variable of integration explicitly at the end of the parentheses.

This is all very elegant and convenient, but how do we know that the result is correct? Taking the derivative of F by the commands

```
diff(F,x); simple(ans)
```

we obtain

```
-1/(a^2*x-x^3)
```

This expression is obviously equal to the function f.

Definite Integrals

We may also evaluate integrals over a given range. As an example of this, let us evaluate

$$\int_a^b \frac{\sqrt{x}}{1+x} dx$$

by the command line

```
clear, syms x a b, f=sqrt(x)/(1+x); Fab=int(f, a, b), pretty(Fab)
```

which gives us the answer

```
   1/2              1/2        1/2              1/2
2 b    - 2 atan(b    ) - 2 a    + 2 atan(a    )
```

Even if we work in the symbolic mode of the program, we may obtain numerical values. For instance, we may evaluate

$$\int_0^{10} \frac{\sqrt{x}}{1+x} dx$$

in a similar fashion, using *vpa* to generate an answer to 14 digits.

```
clear, syms x, f=sqrt(x)/(1+x); Fab=int(f, 0, 10), vpa(Fab,14)
```

The result becomes

```
Fab =
2*10^(1/2)-2*atan(10^(1/2))
ans =
3.7955174050864
```

The limits of integration may even be infinite. For instance, we may calculate

$$\int_0^{+\infty} \exp(-x^2)\, dx$$

by the commands

```
clear, syms x, f= exp(-x^2); Finf= int(f, 0, +inf); vpa(Finf,14)
```

For this definite integral we obtain the value

```
.88622692545275
```

which we recognize from *ex154* (p.97).

An Easy Way to the Simpson Formula

Let us finish this chapter by a practical application. On p.96 we sketched a derivation of the integral under the Simpson parabola. We may now use symbolic methods to obtain this crucial result.

In the following file we first fit a general 2nd-order polynomial

$$f(x) = Ax^2 + Bx + C$$

to the points (x_1, f_1), $(x_1 + h, f_2)$, $(x_1 + 2h, f_3)$ and solve a system of linear equations to obtain the coefficients A, B and C. Finally, we integrate the resulting polynomial from x_1 to $x_1 + 2h$.

```
% ex162.m: Simpson Formula
clear, echo off, syms x x1 h f1 f2 f3 A B C
eq1=A*x1^2+B*x1+C-f1;
eq2=A*(x1+h)^2+B*(x1+h)+C-f2;
eq3=A*(x1+2*h)^2+B*(x1+2*h)+C-f3;
[A0 B0 C0]=solve(eq1,eq2,eq3, A,B,C);
integral=int(A0*x^2+B0*x+C0, x, x1, x1+2*h);
area=simplify(integral)
```

The final answer becomes

```
area =
1/3*h*(f1+4*f2+f3)
```

as we expected. It may be instructive to remove semicolons to view the intermediate results as well.

Exercises

❑ Use symbolic methods to evaluate $\lim_{x \to +\infty} x \sin\left(\frac{1}{x}\right)$.

❑ Use symbolic methods to evaluate $\lim_{x \to +\infty} \left(1 + \sin\frac{a}{x}\right)^x$.

❑ Evaluate $\lim_{x \to 0+} x^{\sqrt{x}}$.

❑ Expand exp(x) in a Taylor series to 6th order around x=0.

❑ Expand exp(x) in a Taylor series to 6th order around x=1. Use *vpa* to truncate the coefficients to 4 digits.

❑ Find an expression for $\sum_{m=1}^{n} m^2$ and simplify as far as possible.

❑ Find the primitive function of 1/cos(x) and check the result.

❑ Integrate $\sqrt{\tan(x)}$ and verify that the derivative of the answer is in fact equal to the original function. First form the difference of the two functions to be compared, then apply *simple*.

❑ Evaluate the definite integral of $\sin(x)^6$ over the range $0<x<\pi$.

❑ Evaluate the generalized integral

$$\int_{-\infty}^{+\infty} \frac{1}{\sigma\sqrt{2\pi}} \exp\left[-(y-y_m)^2/(2\sigma^2)\right]dy$$

First declare the real variables by syms y ym s real and then the constant by syms pi.

17 Random Events and Statistics

Earlier on (p.32) we used the function *rand* as a convenient tool for generating large matrices. Now we shall study random number sequences in more detail and use them to simulate real events.

Seeds

A random number in the interval 0…1 appears on the screen in response to the command

```
rand
```

How did MATLAB generate it, and are such answers really random? The truth is that the program calculates these numbers by a mathematical procedure, and hence they are completely predictable. The sequence is so long, however, that it never repeats itself in practical work.

If you generate a sequence by typing

```
format long, rand(1,3)
```

you will find that repeating this command after re-starting MATLAB yields the same numbers. The calculation of a sequence has to start somewhere, and the initial number is known as the *seed*. Whenever you start a MATLAB session, this seed number is invariably the same. You can change it, however, by a command such as

```
rand('seed',13)
```

which sets up a new sequence at 13, without producing any number. In some situations it is of interest to have complete control of the seed numbers, but in this book we shall only be using numbers as unpredictable as possible. Hence we shall begin all script files by

```
rand('seed', sum(100*clock))
```

which sets the seed by the computer clock every time we run a file. You may view the various seed numbers by repeating the command

```
sum(100*clock)
```

In practice, we should hence never have to worry about predictable number sequences.

Randomness

Even with our precautions about seeds you might still be skeptical about the random nature of a *rand* sequence. Let us make some tests.

A simple verification would be to plot two sequences, considering one of them to be *x*-values, the other one *y*-values. If the numbers should not cover the whole range $0 \le x \le 1$, or if the numbers should be interdependent, we would expect this to be visible in the point pattern. The script file required is

```
% ex171.m:  Random Numbers in (x,y)
clear,  echo off
rand('seed',sum(100*clock))                    % Reset random sequence
n=1000;
X=rand(1,n);  Y=rand(1,n);                     % Random row vectors
figure(1),  plot(X,Y, 'o'),  grid on,  xlabel('random x'),  ylabel('random y')
```

where the string argument in the *plot* parenthesis means “blue circles”. A run should generate a plot similar to the one below, but your figure will probably look different in its details. Again, a second run would yield another variant. None of them, however, should show traces of regularity. Evidently, the ranges in both directions are also as expected.

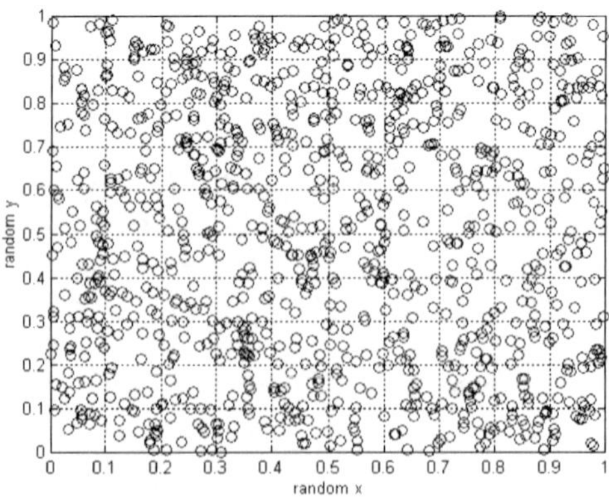

We are by no means limited to the range $0 \le x \le 1$. If we prefer a symmetrical range, $-1 \le x \le +1$ say, we may modify the file by typing

```
X=2*(rand(1,n) - 0.5);  Y=2*(rand(1,n) - 0.5);
```

which becomes obvious when you start thinking about it.

Test by a Histogram

We could estimate the probability of finding points within each sub-interval (above) by counting the circles in each rectangle. MATLAB offers a convenient graphical tool, however, i.e. the *histogram*.

In order to exploit this alternative feature, we only need to type the following file.

```
% ex172.m:  Histogram of symmetric random numbers
clear,  echo off,  rand('seed',sum(100*clock))
n=5000;
X=2*(rand(1,n)-0.5);                          % Vector with n elements
figure(1),  hist(X,10);  grid on              % 10 containers
  title('Number of Random Values in Each Container')
```

Here we generate a symmetric distribution of x-values and present it by a histogram, which sorts the values in the input vector X over 10 intervals (containers) and displays bars representing the *number of values* falling within each container. The function *hist* performs this sorting for us.

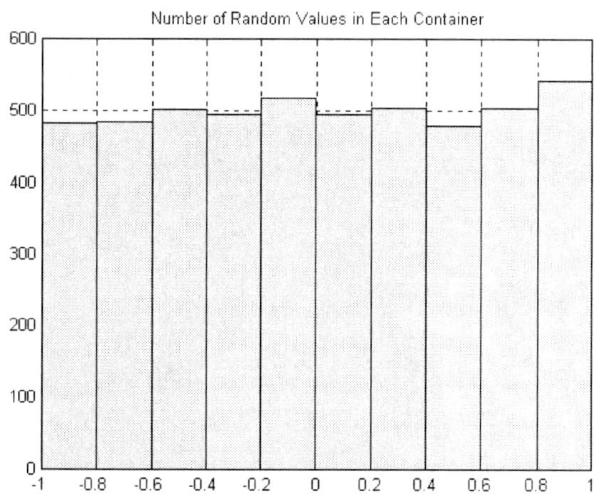

As seen from the above figure, which results from this run, each interval does not contain exactly the same number of random values. This is entirely in accord with statistical theory, as you will gather from your textbook. If we were to increase the number of values n, however, we would obtain a more uniform distribution. Changing the number of containers to 20 results in a more irregular graph, caused by a smaller number of events in each container.

Flipping for Heads or Tails

Flipping a coin is a popular way of illustrating how probability theory works. Although probability models certainly apply to this process, it is tedious to count and plot results of such tests.

MATLAB provides results equivalent to "head and tail" experiments. The following script file lets us perform an unlimited number of "tosses". The command *rand* yields a single random number (scalar), with uniform probability in the interval $0 \leq x \leq 1$. Subtracting 0.5 from each number, we arrive at a distribution from -0.5 to 0.5. We interpret a positive response as a "head" and a negative one as a "tail". The decimals are irrelevant, however, and hence we round off the random numbers to the nearest integer in the direction of $+\infty$. A MATLAB function that rounds off numbers in this manner is *ceil* ("ceiling"). The answer will then be 0 for all negative values and 1 for the others. We may interrupt the sequence by *<Ctrl>c*.

```
% ex173.m:  Tossing Coins
clear,  echo off,  rand('seed',sum(100*clock))
while 1==1                          % Turn forever
  head=ceil(rand-0.5)               % Either 1 or 0
  pause                             % Stop until <Return>
end
```

Running a number of tosses you discover how convenient it is to explore random events in this way.

Repeated Sequences of Tosses

Now suppose we make many test sequences, each involving 10 tosses, say. How many heads do we record during each test? We expect about half on the average, i.e. 5 in this case, but we have also learnt to accept statistical scatter. Repeating the same test we should obtain a distribution of the number of heads.

We could use the preceding file for such experiments, but there is an even more convenient method. If we write the line

```
head=ceil( rand(1,10)- 0.5)
```

we receive ten answers at once. This is already an advantage, but we need not even bother to count the "1s" in the multiple answer on the screen. The command *sum* adds the vector elements, and hence we may just type

```
heads=sum( ceil( rand(1,10)- 0.5))
```

to obtain the number of heads in one test.

Proceeding one step further, we store the recorded number of heads in the elements of a vector, which we then present by a histogram. The following script file does all that work.

For each turn of the loop we record the number of heads obtained in that sequence of tosses. The final result of all the loops is thus a set of ntest numbers, indicating the number of heads found. If one of these numbers is 4, say, the container centered on 4 stores one more event. The *hist* function sorts all of the results into containers.

```
% ex174.m:  Repeated Sequences of Tosses
clear,  echo off,  rand('seed',sum(100*clock))
ntest=1000;                                  % Number of tests
m0=10;                                       % Number of tosses per test
for i=1:ntest
  Mh(i)=sum( ceil( rand(1,m0)-0.5));         % Number of heads counted
end
Xh=0:m0;                                     % Select container centers
figure(1),  hist(Mh,Xh),  grid on
  xlabel('number of heads'),  ylabel('number of results in container')
```

In principle, we should be able to use the command *hist* as before, but in this case we prefer to have the containers centered on integral values. We achieve this by the second argument, Xh, a vector specifying the container centers.

A run yields the histogram below. There are large values at about 5 heads, as expected, but the distribution covers all possible outcomes of the test.

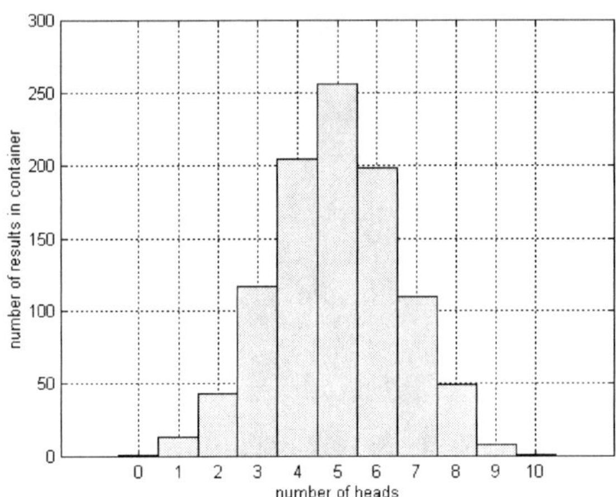

Comparison with Probability Theory

The above problem may be analyzed by combinatorial methods, and the result is the general *binomial distribution*

$$\Pr(m) = \frac{m_0!}{m!(m_0 - m)!} p^m q^{(m_0 - m)}$$ ●

where $\Pr(m)$ is the *probability* of recording m heads in a series of m_0 flips of a coin. The quantities p and q are the probabilities of finding one or the other of the two results. In the present case, the probabilities of heads and tails are equal, i.e. $p = q = 0.5$. This permits us to simplify the expression by putting $p^m q^{(m_0 - m)} = p^{m_0}$.

We may compare this fundamental prediction to our "tossing" experiment by including points corresponding to the above expression in the histogram.

MATLAB calculates the factorial $m!$ by a *product* of the elements of a vector, e.g.

```
prod(1:m)
```

as you may readily test by using some small number for m. Hence, we may code the above formula and plot it. The first part of the script file will be the same as *ex174*, and we only add a few lines at the end.

```
% ex174a.m:  Repeated Sequences of Tosses, Comparison
...
for j=1:m0+1                                  % Vector index
  m=j-1;                                      % Range m=0...m0
  Xm(j)=m;
  Ybd(j)=ntest* prod(1:m0)/ prod(1:m)/ prod(1:(m0-m))*0.5^m0;
end
hold on                                       % Superimpose points (o)
figure(1),  plot(Xm,Ybd,'ro'),  grid on,  hold off
```

In the above loop the index j must be chosen larger than zero, since MATLAB expects positive vector indices. Hence, we must subtract unity to obtain the number m for the theoretical expression. The latter yields the *probability* $\Pr(m)$ of obtaining a given number of heads, so we must multiply that by ntest to convert to numbers.

We plot the binomial distribution by red circles ('ro'). The new figure demonstrates that the theoretical result (o) is in reasonable agreement with the head count.

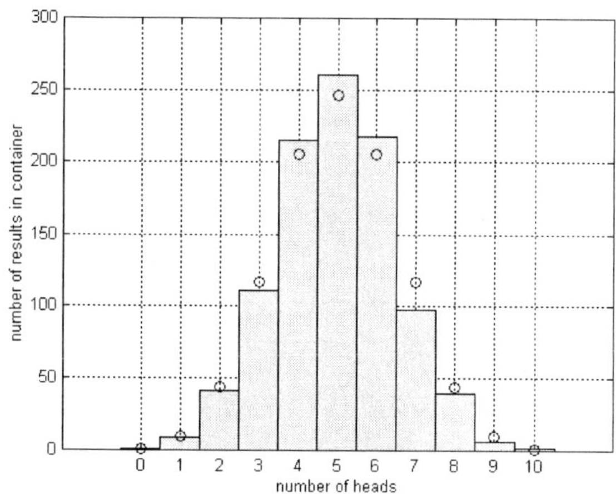

So far we have only made 10 flips of the coin per test. Let us now change to the value m0=100 by the following modifications of *ex174a*.

```
% ex174b.m:  Repeated Sequences of Tosses
...
m0=100;                                    % Number of tosses per test
...
figure(1),  plot(Xm,Ybd,'k'),  grid on,  axis([0 100 0 90]),  hold off
```

In the figure below we notice that the distribution has become much more strongly centered on the average number of heads. The chances of recording as few as 20 heads, say, seems to be negligible.

The binomial distribution, presented by a full curve, now describes the empirical data even better than before.

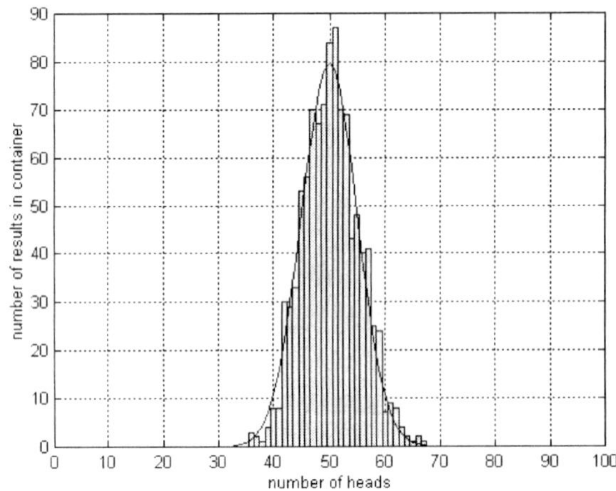

Mean Value and Standard Deviation

The mean value of a sequence of n measurements y_i is defined to be

$$< y > \equiv \frac{1}{n}\sum_{i=1}^{n} y_i$$

Just after running *ex174b* the variables are still available for you to calculate the mean value of the number of heads in rounds of 100 flips.

```
mh_mean=sum(Mh)/ntest
```

Repeating *ex174b* followed by this line a few times you will get different answers, but all of them will be close to 50. This is what we expect, since there are 100 tosses and half of them should yield heads on the average.

The *standard deviation* is a measure of the scatter of a variable y around the mean value and is defined by

$$\sigma = \sqrt{\frac{\sum_{i=1}^{n}(y_i - <y>)^2}{n-1}}$$

In the tossing rounds the mean value of the number of heads was $m_0/2$, and the line required for calculating the standard deviation becomes

```
sdev=sqrt( sum( (Mh-m0/2).^2)/(ntest-1))
```

which you also may execute after *ex174b*. Textbooks show that the mean value and the standard deviation of the binomial distribution may be written

$$<m> = m_0 p$$
$$\sigma = \sqrt{m_0 pq}$$

or $\sigma = \sqrt{m_0}/2$ in the present case. It is interesting to compare this estimate to the known results of the tossing game. The theoretical value is $\sigma = \sqrt{m_0}/2 = 5$, while the numerical data yield something in the range 5 ± 0.1.

The tossing of coins is only a particular situation where random errors play a role. Random errors are relevant to any experimental measurements of scientific quantities, even at the highest level of precision. In such cases there is generally no theoretical basis for assessing “random” errors, but they may always be estimated by repeated measurements.

Binomial and Normal Distributions

In cases where the standard deviation has been determined by practical observation we may still be guided by the binomial distribution, using large numbers for *m* and *n*.

It is an empirical fact that the *normal distribution*

$$\mathrm{Nd}(y) = \frac{1}{\sigma\sqrt{2\pi}} \exp\left[-(y - <y>)^2 / (2\sigma^2)\right] \qquad \bullet$$

describes the scatter of results in most types of measurements. The constants in this expression are such that the integral over *y* becomes equal to 1, as in the case of the binomial distribution.

Let us now convince ourselves that the normal distribution approximates the binomial distribution for a large value of m_0, e.g. $m_0 = 100$. To do this we substitute the parameters $<m>$ and σ for the *binomial* distribution in the expression for the normal distribution.

```
% ex175.m:  Binomial and Normal Distributions
clear,  echo off,  close all
m0=100;                                         % Number of tosses per test
sdev=sqrt(m0)/2;                                % Standard deviation of Ybd
for j=1:m0+1                                    % Binomial distribution
  m=j-1;
  Xm(j)=m;
  Ybd(j)=prod(1:m0)/prod(1:m)/prod(1:(m0-m))*0.5^m0;    % Binomial
  Nd(j)=1/sdev/sqrt(2*pi)* exp(-(m- m0/2)^2/2/sdev^2);     % Normal
end
figure(1),  plot(Xm,Ybd,'y',  Xm,Nd,'k--'),  grid on,  axis([0 100 0 0.09])
  title('Binomial and Normal Distributions')
  xlabel('number of heads'),  ylabel('probability')
figure(2),  plot(Xm,Ybd-Nd),  grid on,  axis([0 100 0 3e-4]);
```

The figure below shows the first plot. The two curves merge completely in this plot, but since the dashed curve for the normal distribution Nd overlays the yellow one, both curves can be seen.

The second plot presents the difference (not shown here) and reveals that this is about 1e-4 in regions on either side of the peak. It is clear from these plots that the normal distribution can replace the binomial distribution for most practical purposes.

In this example, the standard deviation is exactly sdev=5. For any random sequence of values, this quantity is of great importance for expressing the amount of scatter. As we can see from the last figure, very few values occur outside a range of $\pm 3\sigma$.

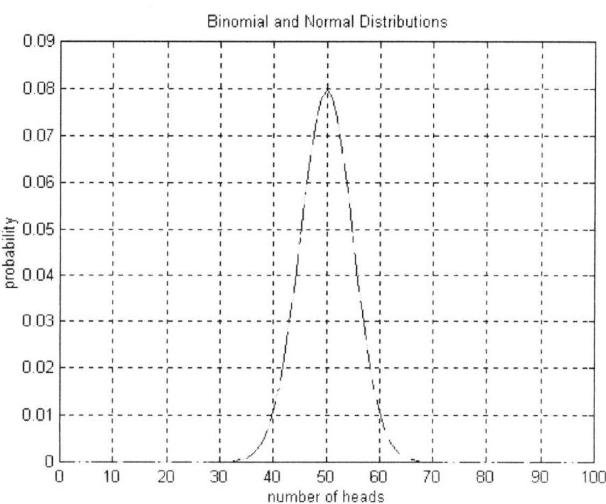

Exercises

❑ Modify *ex172* by changing the number of containers to 20, 40, and 80. What is your qualitative conclusion?
❑ Investigate numerically if Pr(m) is symmetrical with respect to $m0/2$ if p=q=1/2. Also show that the sum of Pr(m) over all m is unity.
❑ Heads and tails are equally probable only if the coin is symmetric. If the tail side of the coin were made of a heavier metal, heads would become more probable. Modify *ex174* to make the occurrence of heads twice as probable.
❑ Test the preceding results against the binomial distribution.
❑ Simulate the random walk of a drunken sailor in the (x, y) plane, starting from a lamppost at the origin. Assume a constant step of unit length and randomize the direction angle for each step. Plot the first 100 steps.

18 Fitting a Function to Data

From the preceding section we know that the *standard deviation* is a measure of the scatter of points around a mean value. We also know how to compute the mean value $< y >$, but let us now define this value in a more general way.

Mean Value as a Minimum

Let y_i (i=1...n) be a sequence of measurements, and let us define the mean value as the value of the parameter p that *minimizes* the expression

$$\sigma(p) = \sqrt{\frac{\sum_{i=1}^{n} (y_i - p)^2}{n-1}}$$

This is a way of searching for the value of p that is, in a sense, closest to the data points. The function σ depends only on p, and our problem is to find the position p_m for the minimum.

The function file for σ, corresponding to a particular data set, may be written

```
% sdev.m:  Standard Deviation
function f=sdev(p);
Y=[1.5284  -0.1022   1.9483   2.8576   0.5422];
n=length(Y);
f=sqrt( sum( (Y-p).^2 ) /( n-1) );
```

Here, the argument p was chosen to be a scalar. To locate the minimum of σ, we plot this function by the following script file. Since the function file only calculates for one value of p at a time, we have to make a loop to obtain values of the standard deviation corresponding to the elements of the vector P.

```
% ex181.m: Standard Deviation to Minimize
clear,  echo off, close all
P=1:1e-3:2;  np=length(P);            % Number of parameter values
for i=1:np
  F(i)=sdev(P(i));
end
figure(1),  plot(P,F),  grid on,  zoom,  xlabel('p'),  ylabel('sdev')
```

The plot resulting from these files is shown below. Evidently, there is a minimum at about $p_m = 1.35$, and if we zoom by clicking on the deepest part of the curve we may estimate one more decimal.

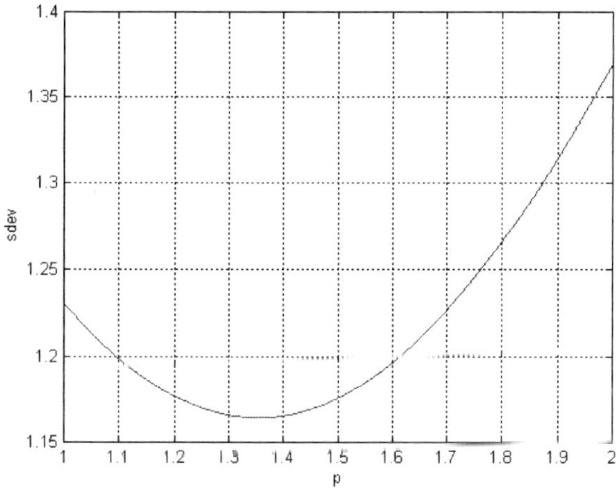

Typing the command line

```
Y=[1.5284  -0.1022   1.9483   2.8576   0.5422];  y_mean=sum(Y)/5
```

according to the simple definition we obtain $< y >= 1.3549$. The alternative definition of the mean value, as the closest fit to data points, is useful in the sense that it may be generalized, as demonstrated in the next section.

We may note that the number n is constant while we minimize $\sigma(p)$. Thus, we could just as well omit the factor $(n-1)$. Also, the square root takes its minimum when thc sum inside it does. Hence, we may simplify further by minimizing the sum

$$S_{sq} = \sum_{i=1}^{n} (y_i - p)^2$$ ●

Elements of Curve Fitting

It often occurs in experimental sciences that one measures a quantity for a number of parameter values. For instance, we might want to measure the electric current through a piece of metal wire at various temperatures, keeping the voltage drop over the specimen constant. The result of a series of such measurements would be a table containing currents I_i in one column and a corresponding set of temperatures T_i in another. Repeating this series of measurements and plotting the data, we might find rather large scatter of the two curves of I against T. This scatter might be caused by difficulties with holding the voltage sufficiently constant during a measurement.

We know that the conductivity of a pure metal should be approximately proportional to $1/T$, T being the absolute temperature. Hence we could assume the relation

$$I = p/T$$

to describe the variation with temperature. The parameter p is so far unknown, but it may be determined by an analysis similar to what we just used for the mean value. This procedure is called *curve fitting*. We will not be able to find a value of p that makes the theoretical relation fit the measured data *exactly*, but there is always a *best* fit.

We might imagine a third column in the table of measured data, containing values of $I = p/T_i$ for the temperatures measured and for a constant trial value of p. The predicted values (I) and those measured (I_i) will differ, and we may calculate the standard deviation (σ) as before. A different value of p will yield a different value σ. The *best* fit is that with the smallest standard deviation. It just remains to determine the value of p that minimizes σ.

The function file for the sum of $(I_i - I)^2$ will be as follows.

```
% ssq1.m:  Sum of Differences Squared
function f=ssq1(p);
global T I
f=sum( (I-p./T).^2 );                    % Sum squared, Ssq
```

In the previous file *sdev* we summed over (Y-p) squared, but in the present problem the values for comparison are $I = p/T_i$, and for this reason we must sum over (I_i-p/T_i) squared.

A new feature with the above file is the *global* statement. Since the function file may be used for different cases, it is not convenient to have it contain data from a specific experiment. Instead we let the vectors T and I share the values within the same vectors in the following script file. In this way, we make the measured vectors available without fixing their contents. When global variables are involved, it is wise to erase all traces of earlier values in both the script and the function files by the new command *clear all.*

```
% ex182.m:  Minimize Standard Deviation
clear all,  echo off, close all
global T I                              % Must come after "clear all"
T=[200 250 300 350 400];
I=[0.1176   0.0926   0.0735   0.0711   0.0594]*1e-7;
P=1.5e-6:5e-9:3.5e-6;  n=length(P);
for i=1:n
  F(i)=ssq1(P(i));
end
figure(1),  plot(P,F),  grid on,  zoom,  xlabel('p'),  ylabel('sum of squares')
```

After running this file we may zoom to locate the value of p that minimizes S_{sq}, and we find approximately $p_m = 2.34e-6$. Substituting this value into the fitting function, we have $I = 2.34e-6/T$. We are now in a position to plot the measured data together with the best-fitting curve. We just add a line to *ex182* as follows.

```
% ex182a.m:  Minimize Standard Deviation, Plot Result
...
figure(2),  plot(T,I,'o',  T,2.34e-6./T),  grid on,  xlabel('temperature')
```

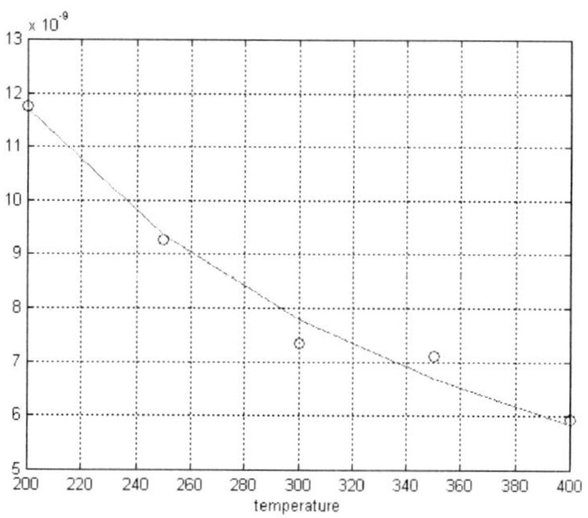

The curve in the above plot clearly is a polygon, because we plotted against the measured temperatures, i.e. only five values.

Determining the Fitting Parameter

To read off the parameter value corresponding to the minimum in S_{sq} is not the most convenient way to obtain results. MATLAB has a built-in routine (p.67) for finding the minimum of a function, i.e. *fmin*. The script file below yields the parameter value p_m automatically.

```
% ex182b.m:  Minimize Sum of Squares by FMIN
clear all,  echo off
global T I
T=[200 250 300 350 400];
I=[0.1176   0.0926   0.0735   0.0711   0.0594]*1e-7;
pm=fmin('ssq1', 0,1, [1 1e-9])
```

The first argument of *fmin* is the name of the function to be minimized, and the next two arguments specify the interval where we search for the value pm. The last argument is a vector containing the requested accuracy as the second element. The first element is 1, which means that we wish to see the intermediate steps in the search for the minimum. The value 0 would mean that we abstain from these details.

The above method of adjusting a function to measured data is usually called *least-squares* fitting, in view of the form of the function ssq1.

Curve Fitting to External Data

In the preceding section we typed the experimental data directly into the script file, but generally we would like to use the latter for several different sets of experimental data.

One solution would be to enter the data from the keyboard, but that is not a very practical procedure. The disadvantage is that a single mistake forces you to repeat the typing from the beginning. Also, you can't add results of further measurement afterwards.

A better way to enter data is to type the values into a data file by means of the MATLAB M-file Editor. For instance, let us type one column for the temperature and a second one for the current, as we would jot them down when taking measurements.

```
% metal.m:  Temperature and Current
200   0.1176e-7
250   0.0926e-7
300   0.0735e-7
350   0.0711e-7
400   0.0594e-7
```

The above is an M-file containing only *data*. Once this file has been saved, we may recall it by the *load* function. For this we need a script file like

```
% ex183.m:  Least-Squares Fit from Data File
clear all,  echo off , format short e
global T I
load('metal.m')                         % Load the data from file
metal                                   % Verify input data
T=metal(:,1);  I=metal(:,2);            % Separate into two columns
pm=fmin('ssq1', 0,1, [1 1e-9])          % Find minimum
n=length(T);  ssq=ssq1(pm);             % Least sum of squares
standard_deviation=sqrt( ssq/(n-1))
format short                            % Return to default format
```

After recalling the data matrix by *load*, we separate it into the two column vectors T and I. The temperatures are in the first column, and in order to extract T we need to specify T=metal(:,1). This means that the vector T is put equal to all (:) rows (elements) in the 1st column, and so on. The *global* declaration makes the vectors available to the function ssq1 as well.

Running this file we find the same final value for pm as before. On the last two lines, we use that to calculate the corresponding sum of differences squared and the standard deviation. The latter seems to be in reasonable agreement with the scatter demonstrated by the plot on p.123.

Simulating Measurements

In order to study curve fitting it is convenient to generate a set of data that is similar to what we would obtain from actual measurements. We may do this by adding simulated experimental errors to the theoretical variation of a quantity.

In the preceding chapter we used the function *rand*, which we found to exhibit a flat distribution. There is another function, *randn*, which obeys the normal distribution (p.118). Let us first generate such a sequence and explore its properties. In the file below we calculate the mean value and the standard deviation for a vector generated by *randn*. Notice that the command containing *seed* is different from what we used before.

```
% ex184.m:  Normal Distribution from Random Numbers
clear all,  echo off
randn('seed', sum(100*clock));
n=10000;  Y=randn(1,n);
ymean=sum(Y)/n                                  % Mean value
standard_deviation=sqrt( sum( (Y- ymean).^2) /(n-1) )
```

Running this script file a few times we find mean values that are very small compared to the standard deviation, which is close to 1. We may thus use *randn* to add noise to any function, sampled at a sequence of points, to give it the aspect of measured data.

Motion of a Ball

We shall now consider the simplest possible experiment. Let us throw a steel ball vertically upward inside a vacuum tank with a glass window, so that we can make measurements of the position of the ball as a function of time. Practically, we might illuminate it by light flashes at constant time intervals and record the images on a photographic plate.

Newton's law of motion gives us the following expression for the height y against time

$$y = v_0 t - \frac{1}{2} g t^2$$

where v_0 is the initial velocity, g the acceleration due to gravity, and t the time. In the experiment proposed we measure a sequence of heights y_i corresponding to the times t_i. We assume that the flashes occur very regularly, making time errors negligible. The following script file calculates a probable result of such a set of measurements.

```
% ex185.m:  Vertical Motion of a Ball, Simulated
clear all,  echo off,  randn('seed',sum(100*clock))
v0=12;                                  % SI units throughout ...
g=9.81;                                 % Acceleration due to gravity
T=0:0.02:2.5;                           % Times as a row vector
Y=v0*T- 0.5*g*T.^2;                     % Theoretical heights
Err=0.3* randn(1,length(T));            % Row vector, the length of T
Y=Y+Err;                                % "Measured" heights
figure(1),  plot(T,Err,'o'),  grid on,  title('Error against time')
figure(2),  plot(T,Y,'+'),  grid on,  title('Height against time')
Ball=[T' Y'];                           % Two columns (transposed rows)
save ball.m  Ball  -ascii;              %  Save data table in \matlab\bin
```

Figure 1 shows the deviation of measured points from the ideal parabola. Figure 2 (below) displays the measured values *versus* time.

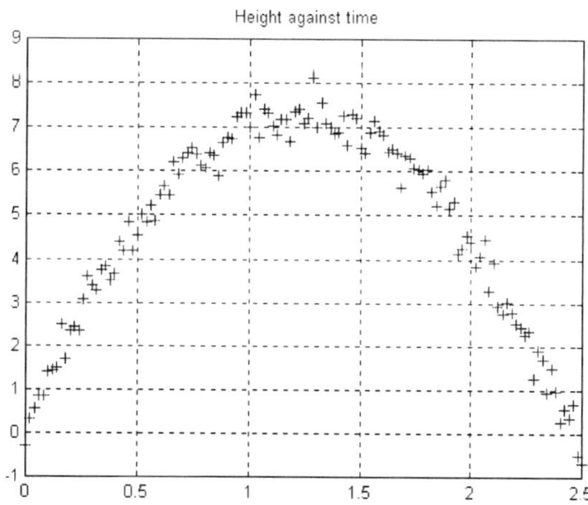

Of course, we could now make use of these data points directly, but the illusion of an experimental measurement will be more complete if we store them in a file. We may later draw from this table of data, as if it had been recorded during real measurements.

For this reason we have prepared the matrix Ball, containing times and heights in two columns. Then we *save* that in *ball.m* and add the name Ball for the matrix to be stored in *ascii* format. By default, MATLAB stores this file in *\matlab\bin*, but you may specify your own path by an insert before ball.m in the above file.

This means that if we later type

```
load ball.m, Ball
```

we retrieve the two columns of measured data. For the curve fitting to follow we need to separate the two columns again by the line

```
T=Ball(:,1), Y=Ball(:,2)
```

as we have done before.

Curve Fitting

We shall now assume that the above heights have been measured at a sequence of times. Our next step is to use a theoretical model to fit a function to the data, i.e.

$$y = v_0 t - (g/2)t^2$$

which follows from Newton's law. Let us pretend that the actual values of v_0 and g are not known, and use the empirical information in *ball.m* to determine these parameters.

We thus assume the following expression for the vertical position

$$y(t, p_1, p_2) = p_1 t - p_2 t^2$$

and we now wish to find the values p_1 and p_2 that make $y(t, p_1, p_2)$ fit the measured height data as well as possible.

As before, we minimize the sum of differences squared, which may now be written

$$S_{sq} = \sum_{i=1}^{n} (y_i - y(t_i, p_1, p_2))^2 \qquad \bullet$$

The difference with respect to the previous problem is that we now have *two* parameters to determine, but we shall see that this is not more difficult. We just need to type a new function file for S_{sq}. This function depends on p_1 and p_2, combined into the vector P.

In principle we could plot $S_{sq}(p_1, p_2)$, which would give us a curved surface above the (p_1, p_2) plane. This surface would have a minimum at the coordinates (p_{m1}, p_{m2}), corresponding to the best-fitting parameters.

```
% ssq2.m:  Sum of Squares for the Motion of a Ball
function f= ssq2(P);
global T  Y
Ym=P(1)*T- P(2)* T.^2;                        % Model for fitting
f=sum( (Y-Ym).^2 );                           % Sum of squares
```

To find the parameters p_{1m} and p_{2m} we need to minimize the function *ssq2*. MATLAB offers a powerful procedure, *fmins*, for locating a minimum of a function depending on several variables. It finds the minimum by an intelligent trial-and-error procedure, increasing the step-length where the search direction appears to be favorable. The usage is somewhat similar to that of *fmin*, as illustrated in the general script file below. The main difference between this file and *ex183* is that we now have two parameters to determine.

```
% ex186.m:  Find Parameters for Ball
clear all;  echo off
global T Y
data_file=input('First name of data file= ','s');   % String input
ssq_file=input('First name of ssq file= ','s');
name=[data_file  '.m'];                     % Assemble the parts of the name
load(name);                                 % Load the data from file
M=eval(data_file);                          % Matrix stored as "data_file"
T=M(:,1);  Y=M(:,2);                        % Separate into two columns
Pm=fmins( ssq_file, [6 6], [0 1e-6])        % Vector of final parameters
n=length(T);  ssqm=feval(ssq_file,Pm);  % Least sum of squares
standard_deviation=sqrt( ssqm/(n-1))
```

Here, we start by inputting two file names from the keyboard. The final argument 's' tells the program that the name entered is a character string. In the case of the data file we need to add the extension *.m*, which we achieve by combining the first and second name into a vector.

When we wish to evoke the data matrix, however, we must first convert the string data_file into a MATLAB command, and that is the role of *eval*.

A similar situation exists with the function *feval*. We normally invoke a function by its name, e.g. sin(x). The variable *ssq_file* is not the name, however, but it *contains* the name. The "manager" function *feval* lets us use functions having any name. A roundabout way of calculating sin(1) would thus be

```
feval('sin',1)
```

All this is a little technical, but there is nothing to be understood: we only apply syntax rules. The advantage of this new script file is that it is rather general. It handles different sets of data and different fitting functions.

When we run the above script file, the data file results in a matrix named M, which is similar to the array that we typed into the Editor. The temperatures are in the first column, and in order to extract T we need to specify T=M(:,1), and so on.

The first argument of *fmins* is the name of the function to be minimized (ssq2). The next one is a vector containing approximate values of p_{1m} and p_{2m}. The final argument is a vector containing the maximum tolerable error in the parameters.

Running this file with the input strings *ball* and *ssq2* we might obtain $Pm(1) = 11.9959$ and $Pm(2) = 4.8952$, which means that the acceleration would be twice that, i.e. 9.7904. Thus we do not retrieve exactly the values we originally used to create the data file. This is not due to the error in *fmins*, which is only 1e-6, but it arises from the random errors we added to the data. If we repeat *ex185* to generate a new set of data, we indeed obtain a different result for Pm.

The last two command lines evaluate the standard deviation corresponding to Pm. Notice that this value agrees well with the scattering introduced by means of *ex185* and shown on p.127.

Residual after the Fit

How can we verify that the function really fits the data? A very direct way would be to plot the final function y, corresponding to the parameters obtained, and display the data in the same frame. An alternative is to plot the deviation (*residual*) of the function from the data points and to estimate its standard deviation. We can do both of these things by adding the following lines to *ex186*.

```
% ex186a.m: Find Parameters and Residual after the Fit
...
F=Pm(1)*T- Pm(2)* T.^2;                      % Function for best fit
figure(1), plot( T,F,  T,Y, 'o'), grid on    % Function and data points
Residual=F-Y;                                % Deviation of F from data
figure(2), plot( T, Residual, '*'), grid on, ylabel('residual')
```

The following figure shows the residual of the fit as a function of time. The span of the residual values agrees reasonably well with the artificial errors we introduced in *ex185*.

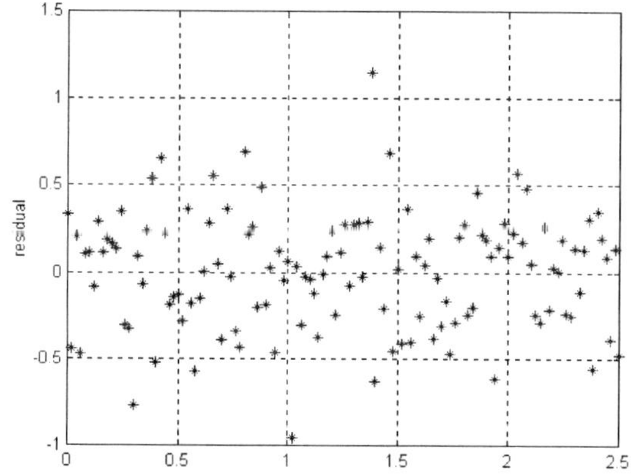

Fitting to Real Experimental Data

In a real-world situation, we would record our data in a file similar to *ball.m*. The computer might even read these values directly from the instruments and produce such a file. The script file for the analysis would thus be analogous to *ex186*. The difference lies mainly in the expression leading to the vector Ym that appears in *ssq2*. The choice of a mathematical model for the process was easy in the case of the ball being thrown vertically, since we had just added random errors to the model function.

In a real case, theory may give a lead to the form of the function to try. If the volume of a gas is measured as a function of pressure, a suitable function could be

$$V = p_1\, p^{-p_2}$$

where V is the volume and p to the pressure, or in MATLAB form

```
V= P(1)*Pr.^(-P(2))
```

The quantity measured may depend on more than two variables. The volume, for instance, generally would be a function of both pressure and temperature (T). A possible model for this more general case would be

$$V = p_1\, p^{-p_2}\, T$$

or for the MATLAB file

```
V= P(1) *Pr^(-P(2)).*T
```

The *fmins* routine can fit a model function to the data, even if there are several measured variables and more than two parameters.

Exercises

❑ Modify *ex185* to reduce the experimental errors by a factor of 100. Run *ex186* again and notice the difference.

❑ Simulate measured data using the function $y = p_1 + p_2 t^2$ with suitable coefficients. Modify *ssq2* accordingly. Run *ex186* to fit this function to the data.

❑ Modify *ex185* to simulate about 50 measurements of the function $y = 100x\exp(-2x)$ over the range $0 < x < 3$. Adapt *ssq2* to that function and then fit to the data using *ex186*. Notice that *ex186a* refers to the *global* vectors T and Y. Hence, *ssq2* must include the assignment X=T after the global statement.

❑ Simulate a set of measurements of $y_i = 2\sin(t_i) + 5\cos(t_i)$ over a suitable interval of t and then fit the function $y = p_1 \sin(t) + p_2 \cos(t)$ to the data.

❑ Use $y_i = 3\sin(x_i) + 2\exp(-x_i)$ to simulate measurements, and then try to fit the *different* function $y = p_1 \cos(x) + p_2 \exp(-x)$ to the data.

❑ Run *ex185* to obtain simulated data and fit by *ex186*. After the model of *ex181*, first make a plot of ssq2 versus p1, keeping p2=Pm(2). Then modify *ex186* again to plot against p2, this time keeping p1=Pm(1).

19 First-Order Differential Equations

An ordinary differential equation, abbreviated ODE, relates derivatives of a function to the function itself. A general ODE of first order may be written

$$\frac{dy}{dx} = f(x, y)$$

where f is a known function. In cases of practical interest there is one, and only one, solution $y(x)$ through any given point (x_1, y_1).

Differential equations are of crucial importance in science and technology. They may be said to summarize empirical and analytical knowledge in these fields. They express how quantities change with space coordinates or time, given certain initial values. These initial values have to be determined experimentally.

One example is the decay of a radioactive species, for which we may predict the amount remaining *versus* time on the basis of the initial amount and a characteristic constant.

Exact solutions exist for many ODEs, and in most cases numerical methods are successful. Before we begin to study an elementary solution algorithm, we shall illustrate graphically what a solution means.

Direction Fields

If we represent a solution to the above ODE by a curve, the tangent at each point must have a slope equal to $f(x, y)$. Even before we have obtained a solution, we may consider $f(x, y)$ as defining slopes at all relevant points in the (x, y) plane, in other words a field of directions. If we indicate the directions by short line segments, we may glean the general behavior of possible solutions.

Let us first take an extremely simple example, i.e. $f(x,y) \equiv -y$. In the following script file we define f by an inline function. Although the function does not depend on X, we include it in the argument parentheses for consistency with future files. Then we proceed to draw the corresponding direction field.

```
% ex191.m:  Direction Field for dy/dx=-y
clear all,  echo off,  close all
F=inline('-Y', 'X', 'Y');
xmax=4;  delx=xmax/30;
ymax=3;  dely=ymax/30;
[Xm,Ym]=meshgrid(0:delx:xmax, 0:dely:ymax);
Dx=Xm-Xm+1;                        % Unit x-components of arrows
Dy=F(Xm,Ym);                       % Compute y-components
L=sqrt(Dx.^2+ Dy.^2);              % Initial length of [Dx Dy]
L=L+1e-10;                         % Add small number to avoid L=0
Dx1=Dx./L;  Dy1=Dy./L;             % Unit length for all arrows
figure(1),  quiver(Xm,Ym,  Dx1,Dy1,'.'),  grid on
  axis([0 xmax 0 ymax]),  title('Direction Field for dy/dx=-y')
```

In the beginning of this file we specify the maximum extent of the field and the intervals between the line segments (arrows). The *meshgrid* function provides the x and y values at all arrow positions for later use. You might test this by typing Xm, say, after a run.

Our purpose is to indicate slopes by short line segments, and initially we just choose $\Delta x = 1$ and calculate the corresponding vertical increment by $\Delta y = f(x,y)\,\Delta x$. The matrix Dx will be of the same size (number of elements) as Xm but will contain only 1s. After the first step of the calculation the line segments thus will have different lengths, which is the reason why we later divide by L to convert to a standard length (unity).

Finally, we exploit the graphical function *quiver*, which literally is loaded with arrows. To make it produce arrows *without* tips we have to add a modifying string ('.') as the last argument. The *axis* command is not strictly necessary, since *quiver* normally plots over the entire field, but this addition eliminates the margin included for the last lines toward large x and y.

Running the file produces the following figure. You should print it using *File, Print* (in the upper left corner of the figure) to obtain a larger version for graphical exercises.

Now try to draw solutions by pencil on the field map. Starting at different points in the (x,y) plane, e.g. (0,1) and (0,3), you will find distinct solution curves. You will soon realize that each initial point defines its own solution. If you start in the interior of the diagram, you may also work both ways to produce a solution. There is only one solution, however, through any given point.

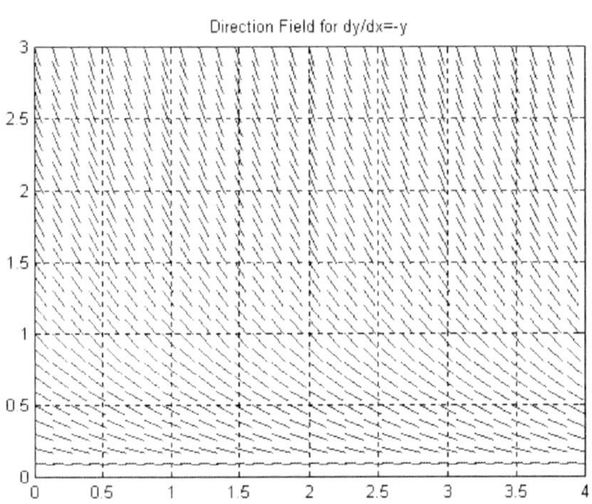

Simple Numerical Solutions

There is an immediate way of solving a first-order ODE, i.e. by the Euler algorithm. Its central idea is to rewrite the ODE in terms of finite differences:

$\Delta y = f(x,y)\Delta x$.

We begin by choosing a constant interval length $\Delta x = h$. From the initial point (x_1, y_1) we find the next y-value on the solution curve $y_2 = y_1 + \Delta y$ by taking Δy from the above relation. The following point will thus be at $(x_1 + h,\ y_1 + f(x_1, y_1)h)$. Continuing from this point we calculate the next one in a similar manner, and so on. The actual solution involves only the statements between *for* and *end* in the file below.

```
% ex192.m:  Euler's Numerical Solution for dy/dx=-y
clear all,  echo off,  close all
F=inline('-Y', 'X', 'Y');
X(1)=input('x1= ');  Y(1)=input('y1= ');
n=400;  h=0.01;
for i=1:n
  dy= F( X(i), Y(i))*h;
  X(i+1)=X(i)+h;
  Y(i+1)=Y(i)+dy;
end
figure(1),  plot(X,Y, 'r'),  grid on,  axis([0 n*h 0 3])
```

Here we enter the start values x1 and y1 from the keyboard and store the calculated points successively in the vectors X and Y. Runs with various combinations of x1 and y1 will yield curves in reasonable agreement with the above direction field.

If you already are a little familiar with differential equations you have probably noticed that the *exact* solution must be of the form $y = c\exp(-x)$, or in terms of the initial values

$$y = y_1 \exp(x_1 - x)$$

We may compare our approximate solution to the exact one by including the following lines at the end of *ex192*.

```
% ex192a.m:  Euler's Numerical Solution for dy/dx=-y, Comparison
...
Y_ex=Y(1)*exp(X(1)-X);
figure(2),  plot(X, Y-Y_ex),  grid on,  title('Error')
```

Starting from the point (0,3) we find a maximum error of 6e-3, which would barely be visible in the scale of the first figure.

For the next example we shall choose the following ODE

$$\frac{dy}{dx} + x^2 y = 1 + x^3$$

We may now use previous script files as models for the following one.

```
% ex193.m:  Euler Numerical Solution to y'+x^2*y=1+x^3
clear all,  echo off,  close all,  hold off
F= inline('1+ X.^3- X.^2.*Y', 'X', 'Y');
X(1)=input('x1= ');  Y(1)=input('y1= ');
```

```
n=300;  h=0.01;
for i=1:n
  dy=F( X(i), Y(i))*h;
  X(i+1)=X(i)+h;
  Y(i+1)=Y(i)+dy;
end
figure(1),  plot(X,X, 'b',  X,Y),  grid on,  axis([0 n*h 0 3]),  hold on
  title('Solution to dy/dx+x^2*y=1+x^3')
% Add direction field for comparison
xmax=4;  delx=xmax/30;
ymax=3;  dely=ymax/30;
[Xm,Ym]=meshgrid(0:delx:xmax, 0:dely:ymax);
Dx=Xm-Xm+1;                          % Unit x-components of arrows
Dy=F(Xm,Ym);                         % Compute y-components
L=sqrt(Dx.^2+ Dy.^2);                % Initial length of [Dx Dy]
L=L+1e-10;                           % Add small number to avoid L=0
Dx1=Dx./L;  Dy1=Dy./L;               % Unit length for all arrows
figure(1),  quiver(Xm,Ym,  Dx1,Dy1, 'g.'),  grid on,  hold off
```

The first half of this file is very similar to the one before. Since the line $y = x$ is of some interest in this problem (as a trivial solution), we draw that line and then the solution curve, so that the latter may overwrite the line where they coincide. The second part, which superimposes a direction plot, is essentially the same as in *ex191*.

The following figure is obtained with the initial values (0,2). Other solutions to explore are, for instance, those corresponding to the starting points (0,1), (0,1e-3), (1,0), and (2,0).

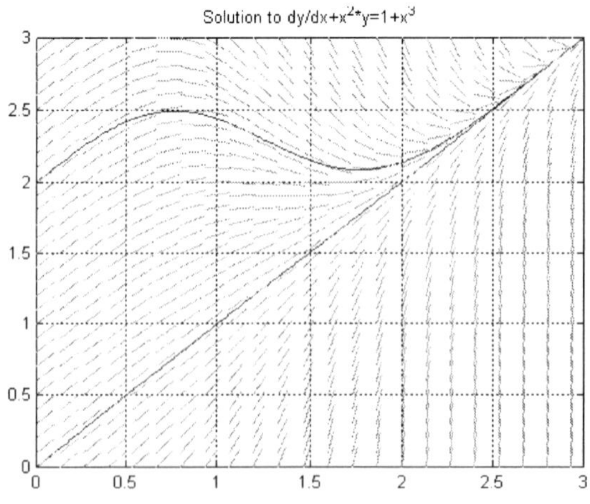

Although the solutions seem to agree well with the direction field, we have not yet assessed the accuracy in detail. Even this ODE has an exact solution, i.e.

$$y = (y_1 - x_1)\exp\left(\frac{x_1^3}{3} - \frac{x^3}{3}\right) + x$$

as you can easily verify by substitution. To compare it to the numerical solution we only need to add two lines to a copy of *ex193*, as follows.

```
% ex193a.m:  Euler's Solution Verified
...
Yex=(Y(1)-X(1)).*exp( X(1).^3/3- X.^3/3)+X;
figure(2),  plot(X,Y-Yex),  grid on,  title('Solution Error')
```

This file yields the plot shown below for the initial values (0,2). The maximum error is about 6e-3, which means that it would hardly be visible in a simultaneous plot of the two solutions. We may improve the numerical result by increasing n and decreasing h by the same factor, say 10. This reduces the error to one tenth, but the file also takes ten times longer time to execute.

Clearly, an exact solution is preferable whenever it can be obtained in reasonably short form. A large class of equations, known as non-linear, can only rarely be solved exactly. Numerical methods usually work, however.

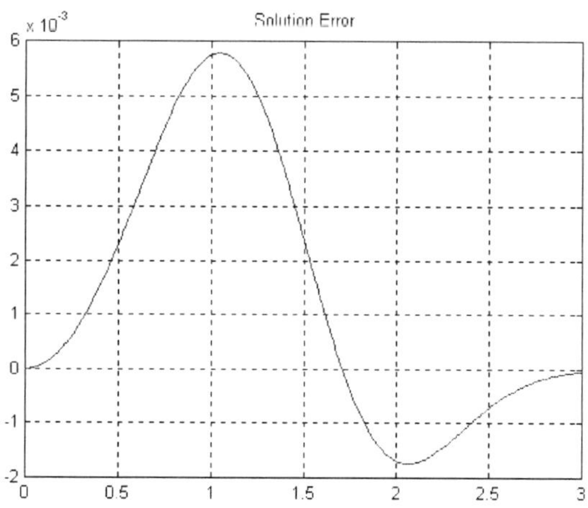

Non-Linear ODEs

An ODE of the form $dy/dx = f(x,y)$ is said to be *linear* (in y) if it may be written in the form $dy/dx = yg(x) + h(x)$. As an example of a *non-linear* ODE we shall take

$$\frac{dy}{dx} = \sin(xy)$$

which presumably cannot be solved exactly.

Let us first have a look at the direction field. You save *ex191* under the new name *ex194* and introduce the new *inline* function sin(X.*Y). In addition you change the variable ranges in *ex194* by putting both xmax and ymax equal to 5. The result below is a much more complex direction field than we have seen before.

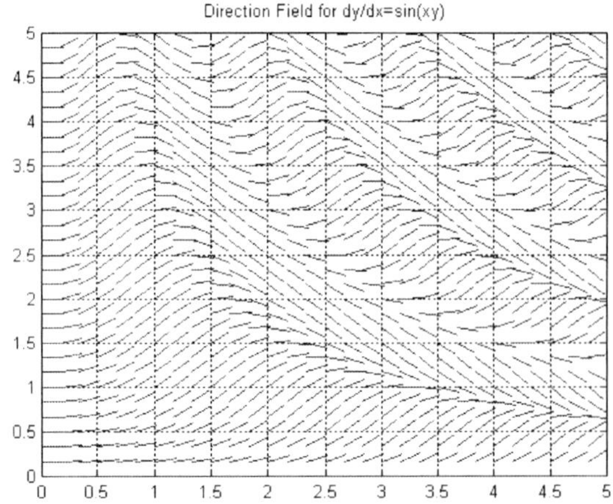

You are now invited to pencil a few typical solutions by joining the line segments judiciously on the printed direction plot. Doing this you will notice that the solutions fall into distinct families, having different behavior toward large x.

This ODE is easily solved by the Euler method. We only need to modify *ex193* to read

```
% ex195.m:  Euler Solution for dy/dx= sin(xy)
clear all,  echo off,  close all,  hold off
F=inline('sin(X.*Y)', 'X', 'Y');
```

```
X(1)=input('x1= ');  Y(1)=input('y1= ');
n=500;  h=0.01;
...
figure(1), plot(X,Y), grid on, axis([0 n*h 0 5])          % First figure(1)
  title( 'dy/dx= sin(xy)'), xlabel('x')          % Eliminate the direction field
```

A run for x1=0 and y1=3 results in the plot shown below. You may also try other initial values.

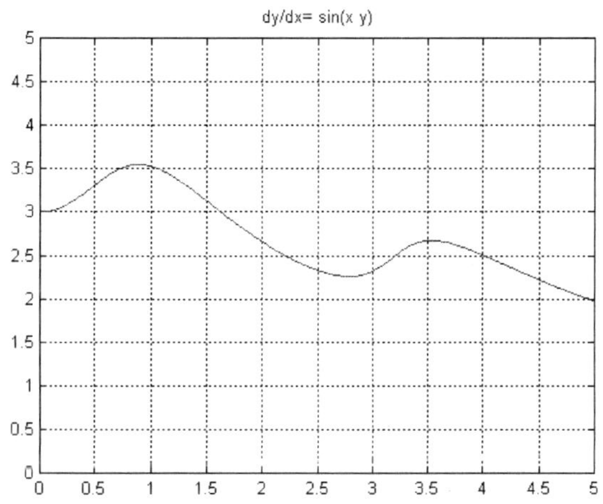

Since the solutions fall into several distinct categories, it would be practical to incorporate more than one solution curve in the same diagram for easy comparison. We may achieve this by *holding* a curve already drawn, as we did on p.102. To obtain several plots in the same diagram we employ an infinite loop, with some facility for stopping when we have collected enough curves.

```
% ex196.m:  Repeated Solutions for dy/dx= sin(xy)
clear all,  echo off,  close all,  hold off
F=inline('sin(X.*Y)', 'X', 'Y');
while 1==1                                   % Beginning of outer loop
  X(1)=input('x1= ');
  if X(1)<0,  break,  end                    % Break on negative input
  Y(1)=input('y1= ');
  n=500;           h=0.01;
  for i=1:n                                  % Beginning of inner loop
    dy=F( X(i), Y(i))*h;
    X(i+1)=X(i)+h;
```

```
    Y(i+1)=Y(i)+dy;
  end                                   % End of inner loop
figure(1), plot(X,Y), grid on, axis([0 5 0 5])
    title( 'dy/dx= sin(xy)'), hold on
end                                     % End of outer loop
hold off
```

The main (outer) loop in the program will stop as soon as you enter a negative value for x1. The figure below gives an example of the set of curves you may generate by this script file.

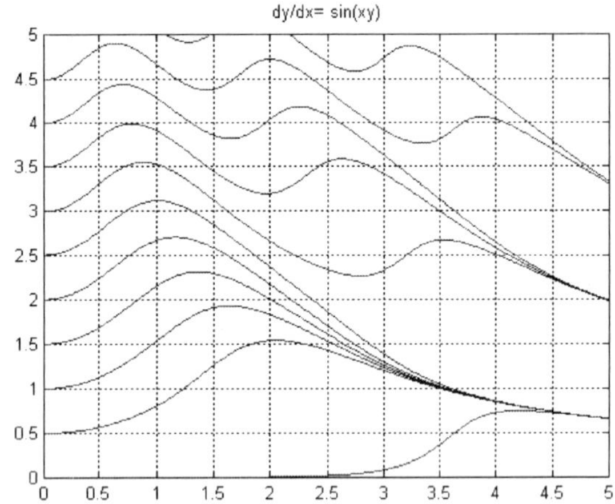

Smarter than Euler

The Euler method allowed us to solve several ODEs using a limited number of basic operations (addition, subtraction, multiplication and division) on the functions appearing in the equation. It only makes use of the definition of the derivative in order to provide an approximate solution. A numerical algorithm could hardly be simpler, and to the author it was a revelation to find that so many reasonably accurate solutions could be generated.

The results we obtained, however, could never be improved by orders of magnitude within realistic calculation times. In order to progress further, we need more sophisticated methods. Fortunately, such tools are included in MATLAB. The routine *ode45* based on the

Runge-Kutta-Fehlberg (RKF) method is both smarter and faster. Its main strategy is to make maximum use of Taylor expansions and adapt the step length to local conditions. The details of the algorithm are outside the scope of this volume.

The RKF method is easy to use. We may modify *ex193a* as follows to solve the ODE by the RKF procedure.

In the *inline* function definition, the output value f corresponding to the derivative y' is a scalar, not a vector as before, which simplifies the syntax since no point operations are necessary. The solution part of the file now condenses into one line.

```
% ex193b.m:  RKF Solution to dy/dx=1+x^3-x^2y
clear all,  echo off,  close all,  hold off
f=inline('1+x^3-x^2*y','x','y');
x1=input('x1= ');  y1=input('y1= ');
[X Y]=ode45(f, [x1 3], y1, odeset('abstol',1e-8,  'reltol',1e-8));
figure(1),  plot(X,X, 'b',  X,Y),  grid on,  axis([x1 3 0 3])
title('dy/dx=1+x^3-x^2y '),  xlabel('x'),  ylabel('RKF solution')
Yex=(Y(1)-X(1)).*exp( X(1).^3/3- X.^3/3)+X;
figure(2),  plot(X,Y-Yex),  grid on,  title('Solution Error'), xlabel('x')
```

Here, the 2nd argument in the *ode45* parentheses is a vector defining the range of x for which we request a solution. The 3rd argument is of course the initial value for y. The 4th argument is a function that sets the maximum error by the two tolerance commands, *abstol* and *reltol*.

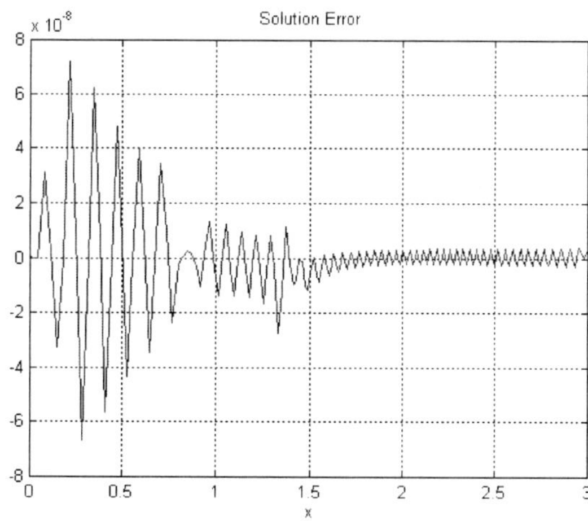

The result of running this file is a solution that appears to be identical to what we obtained by the Euler method. As seen by the above figure, however, the *error* of the solution for the initial values (0,2) is much smaller and the run time is still acceptable.

The error is not as small as requested by *abstol* and *reltol*, however. In addition to this error, we find a much larger fluctuation over the left part of the curve.

Curious Behavior of a Non-Linear ODE

Let us now revisit the non-linear equation $y' = \sin(xy)$ that we already solved by the Euler method. We noticed that the various solutions approach each other asymptotically at large values of x. There are several of these asymptotes, however, and the question is under what conditions the solution switches from one asymptote to the next one. Using the RKF procedure we may quickly investigate how this transition occurs. The following modification of *ex196* is what we need.

```
% ex196a.m:  Repeated Solution of  y'= sin(x y), RKF
clear all,  echo off,  close all,  hold off
f=inline('sin(x*y)','x','y');
while 1==1                    % Beginning of outer loop
  x1=input('x1= ');
  if x1<0,  break,  end        % Break on negative input
  y1=input('y1= ');
  [X,Y]=ode45(f, [x1 8], y1, odeset('abstol',1e-8,  'reltol',1e-8));
  Y( length(Y))                % Display last y-value
figure(1),  plot(X,Y),  grid on,  axis([0 8 0 4])
    title( 'dy/dx= sin(xy)'),  xlabel('x'),  hold on
end                           % End of outer loop
hold off
```

In the script file we display the y-value corresponding to the maximum value of x. This must be the last element in the vector Y, containing length(Y) elements. Using this numerical value as a guide we may easily explore the details of the transition.

From the figure on p.142 we find that the first transition occurs between the initial points (0,2.5) and (0,3), and trials using the above file indicate that the critical condition is close to the second point. In the search for the critical point we may use the method of bisected intervals that we found helpful for finding function zeros (p.62). By this device we locate the transition to a value of y1 between 2.92974 and 2.92975 (with x1=0).

We see that the solution to this ODE is extremely sensitive to the initial coordinates. The behavior illustrated in the figure below is called *bifurcation* or branching. If such an equation were known to govern a scientific process, the predicted results would be practically indeterminate.

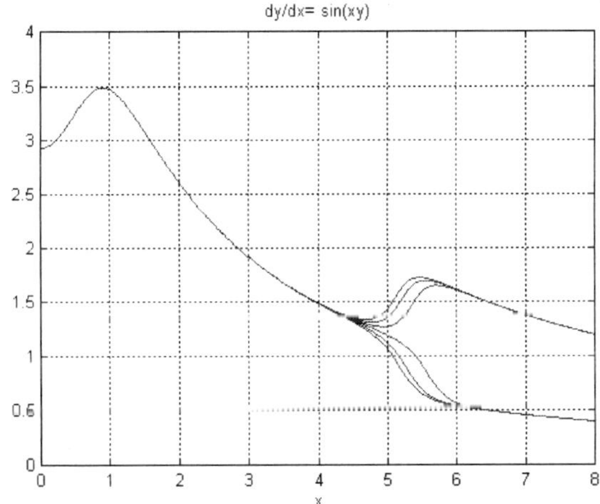

The above phenomenon is related to the concept of *chaos*, which is much discussed these days. The births and deaths of animals of various species may for instance be described by ODEs, and it may occur that predicted populations increase and decrease in what seems to be an erratic fashion, although experiments confirm this behavior.

A more advanced kind of differential equations is used to predict the weather. Under certain conditions, the outcome may depend very sensitively on initial values. An amusing exaggeration of this fact is that a butterfly taking off in India may cause a thunderstorm in the Middle West.

Exercises

- ❑ Modify *ex192* to solve the equation $y' = y\cos(x)$ over the range x=0…10 using the initial value y_1=1 for x_1=0. Plot the solution only. Change the plot commands to accept negative values of y_1 as well.
- ❑ Expand the above exercise by adding the analytic solution Y(1)*exp(-sin(X(1)))*exp(sin(X)) to the file and make a combined plot of the Euler solution and the analytic one. Can you improve the Euler solution?
- ❑ Rearrange *ex193* to show several solution curves in the same plot.
- ❑ Expand *ex196* to display the direction field as well.
- ❑ Judging from the equation $y' = \sin(xy)$ you might expect the derivative to oscillate in the region of large x. Expand *ex196* to plot both xy and $\sin(xy)$ to understand why this is not so. The plots also suggest that the various asymptotes $y_n(x)$ may be obtained from the equation $xy_n = \pi(2n-1)$.
- ❑ Repeat the first two exercises using the RKF procedure.
- ❑ Using *ex193b* as a template, solve $y' + 2xy = 0$ by the RKF procedure over the range x= 0…3. Compare to the analytic solution $y = C\exp(-x^2)$.
- ❑ Include the RKF procedure in *ex196* to compare the solution curves. Name the additional vectors differently, say X1 and Y1. Draw one curve dashed in black, the other one in solid yellow. Why can't you plot the difference in this case?
- ❑ Modify *ex196a* to solve the equation $y' = (y-x)\exp(x-y)$ by the RKF method. Try the initial points (0,1), (0,2), and (0,3). Explain by words why the cutoff occurs.

20 Second-Order ODEs

A general ODE of the 2nd order may be written

$$\frac{d^2y}{dx^2} = f\left(x, y, \frac{dy}{dx}\right)$$

For purposes of numerical solution we may transform it into the following system of equations of the first order

$$\begin{cases} \dfrac{dy}{dx} = y_p \\ \dfrac{dy_p}{dx} = f(x, y, y_p) \end{cases}$$

where the new dependent variable, y_p, is the 1st-order derivative. Although this may not seem plausible at first sight, we may easily use the Euler method to solve such a system of 1st-order ODEs.

For a system of 1st-order ODEs we not only specify an initial function value y_1, but also an initial derivative, y_{p1}. This means that we may treat these two equations in the same manner as the single one before. Knowing the values y_1 and y_{p1} at $x = x_1$ we calculate the next values at $x_1 + h$ by the obvious approximations

$$y_2 = y_1 + \left(\frac{dy}{dx}\right)_{x1} h = y_1 + y_{p1}\, h$$

$$y_{p2} = y_{p1} + \Delta y_p = y_{p1} + \left(\frac{dy_p}{dx}\right)_{x1} h = y_{p1} + f(x, y, y_{p1})\, h$$

Having obtained these new values we proceed in the same way to calculate values at $x = x_1 + 2h$, and so on. In short, the extra dependent variable hardly makes programming more complicated.

Ball suspended by a Spring

As a first example we shall apply Newton's law of motion to an object suspended at the end of a spring. Since time t is now the independent variable, rather than x, we write the equation

$$m\frac{d^2y}{dt^2} = F(y)$$

where m is the mass and F the force, all in SI units (m, kg, s). We disregard the force of gravitation, which is a constant term offset by the elastic force from the spring. If we choose to put the origin, $y = 0$, at the point of equilibrium, the force at a neighboring point y will be $F = -ky$, where k is a constant characteristic of the spring. The ODE thus becomes (friction neglected)

$$\frac{d^2y}{dt^2} = -ky/m \equiv -cy$$ ●

and the corresponding system of 1st order ODEs

$$\begin{cases} \dfrac{dy}{dt} = y_p \\ \dfrac{dy_p}{dt} = -cy \end{cases}$$

We first consider the *function* file required. One of the convenient features of MATLAB is that it permits us to extract more than one vector from a function file. In the present case we want values for the derivatives of y and y_p, which we obtain by

```
% spring.m:  Function for a System of  ODEs
function [dydt, dypdt]= spring(t,y,yp);      % t=time
global c;                                    % Constant from script file
dydt= yp;                                    % dy/dt
dypdt= -c*y;                                 % dyp/dt
```

The global variable c will receive a value from the script file. The two last lines correspond to the system of ODEs.

The *script* file is similar to the ones we already used, the essential difference being the extra line for Yp in the inner loop. The outer loop turns around as long as 1=1, but you may break it by entering zero initial values.

```
% ex201.m:  Euler's solution to y"= -cy
clear all,  echo off,  close all
global c                                    % Value for function file
c= 1;                                       % Value for k/m
n= 1000;          h= 0.01;
while 1==1                                  % Forever true
  T(1)= 0;                                  % Start time, SI units ...
  Y(1)= input('y1= ');                      % Initial position
  Yp(1)= input('yp1= ');                    % Initial velocity
    if  (Y(1)==0) & (Yp(1)==0),  break,  end      % & means AND
  for i= 1: n-1
    T(i+1)= i*h;
    [dydt dypdt]=spring( T(i), Y(i), Yp(i));
    Y(i+1)=Y(i)+ dydt*h;
    Yp(i+1)=Yp(i)+ dypdt*h;                 % Y-prime
  end
figure(1),  plot(T,Y,  T,Yp,':'),  grid on,  title('y"= -cy')
  xlabel('time'),  legend('Y', 'Yp')
end
```

In the above file we enter a position (e.g. 0) and a vertical velocity (e.g. 1.0) from the keyboard. In the real world of physics this means that we take the object to a given height and give it a vertical impact, after which we observe the way it moves.

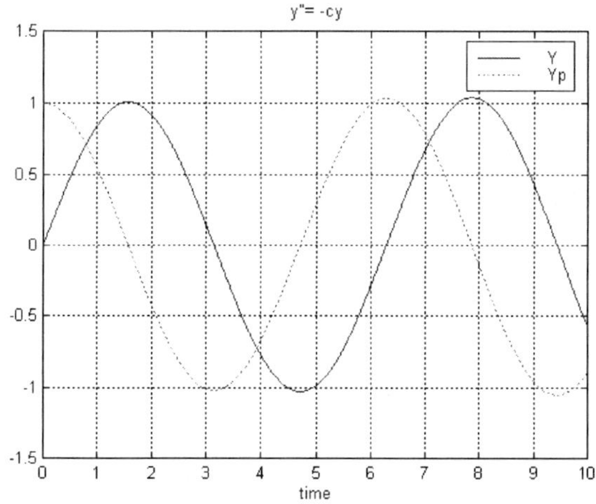

The above figure shows the resulting plot. You may experiment with various input values, positive as well as negative.

Analytic Solution to the Spring Problem

The plots resulting from the above file indicate periodic motion. The current 2nd-order ODE (with $c = 1$) has exact solutions of the form

$y(t) = A\sin t + B\cos t$

where A and B are arbitrary constants. It is easy to show that this function satisfies the ODE. From the conditions $y(0) = y_1$ and $y'(0) = y_{p1}$ we derive an expression of the form we need for comparison, i.e.

$y(t) = y_{p1}\sin t + y_1\cos t$

After including the lines

```
% ex201a.m: Euler's solution to y"= -cy
...
Y_ex=Yp(1)*sin(T)+ Y(1)*cos(T);
figure(2), plot(T,Y,  T,Y_ex,'--'), grid on, title('y"= -cy')
  xlabel('time'), legend('Euler', 'Exact')
```

before the final *end* command in *ex201*, we enter y(1)=0 and yp(1)=1. We then obtain the following graph comparing the Euler solution with the exact one.

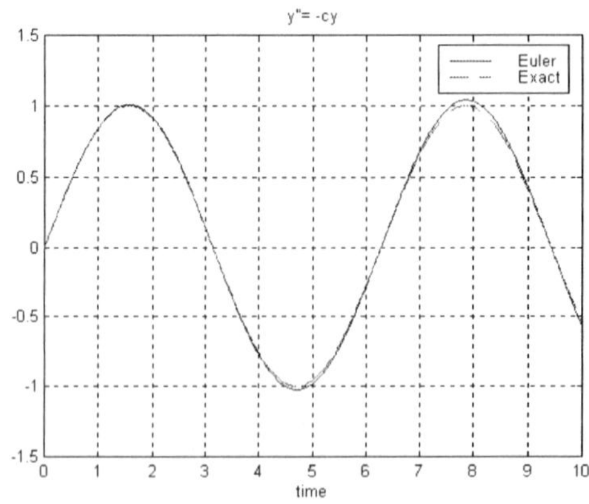

The error becomes clearly visible at large times. Naturally, a numeric solution always involves some error, but we may reduce it considerably by taking a smaller value of h and a correspondingly larger number of points.

Solving a System of ODEs by the RKF Method

The Runge-Kutta-Fehlberg method is equally useful in the case of a system of ODEs. The MATLAB routine *ode45* is designed to accommodate an arbitrary number of equations.

We first have to know how to specify the known function in this case. For a *single* ODE (p.143) we used the simple form

```
f=inline(function, 'x','y');
```

for yp (i.e. y'). With a *system* of two equations, we have two derivatives and an *inline* function can't supply both of them. Hence, we must type a function *file* where the input and output objects are column vectors. Symbolically, we may think of the function file as

$$\begin{bmatrix} dydt \\ dypdt \end{bmatrix} = \text{function}\left(t, \begin{bmatrix} y \\ yp \end{bmatrix} \right)$$

The MATLAB structure of the function will thus be as follows.

```
function Yp=function(t,Y);
```

The first argument of the function is the scalar quantity t, the independent variable. The input column vector Y contains values of the dependent variables, all for time t, and the output column vector Yp the corresponding derivatives.

Swinging Pendulum

An object suspended at the end of a string is a simple form of a pendulum. Textbooks show that the equation of motion is

$$\frac{d^2 y}{dt^2} = -\frac{g}{L}\sin(y)$$ ●

where y now is the *angular* displacement (in radians) from the vertical, g the acceleration due to gravity and L the distance from the fixed point to the center-of-mass of the object. Evidently this ODE is non-linear ($\sin(y)$ being a power series in y). It may be transformed into the following system of 1st-order ODEs, using the notation $c \equiv g/L$.

$$\begin{cases} \dfrac{dy}{dt} = y_p \\ \dfrac{dy_p}{dt} = -c\sin(y) \end{cases}$$

The corresponding function file becomes

```
% pend.m:  RKF Function for Pendulum
function Yp=pend(t,Y);
global c
dydt=Y(2);                              % dy/dt= yp
dypdt=-c*sin(Y(1));                     % dyp/dt= -c sin(y)
Yp=[dydt  dypdt]';                      % Transpose to column
```

The script file required is also very similar to the one before. Let us consider the pendulum in its equilibrium position, with y=Y(1)=0. In this state we strike it sideways to impart an initial angular velocity yp1 and watch the subsequent motion. In a practical case, the initial velocity will be non-zero, so we may use the trivial value yp1=0 to stop the run. The following file applies to the pendulum problem.

```
% ex202.m:  RKF solution to y"= -c sin(y)
clear all,  echo off,  close all
global c                                % Available to pend.m
c= 9.81;                                % Value of g/L for L=1
while 1==1
  y1=0;                                 % Initial angular position
  yp1=input('yp1= ');                   % Initial angular velocity
  if yp1==0,  break,  end
  Y_initial=[y1  yp1]';             % Transpose to column
  options=odeset('abstol',1e-8,  'reltol',1e-8);          % For ode45
  [T,Y]=ode45('pend',  [0 5],  Y_initial, options);
  figure(1),  plot(T,Y),  title('y"= -c sin(y)'),  grid on,  xlabel('time')
end
```

An initial value of yp1=1.0, say, yields a plot with the expected oscillatory behavior.

You may be surprised to find that the angular velocity, Yp, appears in the plot, although we have not expressly requested it. The pendulum equation is a 2nd-order ODE, but *ode45* solves it as a system of 1st-order ODEs. Hence, it yields two solutions, one for y and one for y_p. By displaying the solution Y on the screen you can confirm that this is in fact a matrix containing two column vectors. When the *plot* routine encounters a matrix, it plots one curve for each of the columns.

If you do not want a plot of the angular velocity in the same diagram, you might instead type

```
plot(T, Y(:,1) );
```

to select the first column vector.

Approximate Analytic Solution

Elementary textbooks usually treat the motion of the pendulum by the approximation $\sin(y) \cong y$, yielding the ODE

$$\frac{d^2y}{dt^2} = -cy$$

This ODE is identical to what we used for the object suspended by a spring (p.148). We shall now investigate how well this approximation works for a few values of the initial angular velocity yp1. For this purpose we include the exact solution to the *simplified* ODE

$$y_{ex} = \frac{y_{p1}}{\sqrt{c}}\sin(t\sqrt{c}) + y_1\cos(t\sqrt{c})$$

and plot it along with the RKF solution to the non-linear problem. The following lines need to be inserted *before* the final end.

```
% ex202a.m:  RKF solution to y"= -c sin(y), Compared
...
  a= sqrt(c);  Yex=yp1/a*sin(a*T)+ y1*cos(a*T);
  figure(1),  plot(T,Y(:,1),  T,Yex,'--'),  grid on,
     legend('Y','Yex'),  xlabel('time')
```

At a small initial velocity, say yp1=0.01, you will find good agreement with the approximate solution. At a larger value, e.g. yp1=1.0, we notice increasing discrepancy between the two curves for y as time increases, and it becomes considerable in the right half of the plot. Inputting yp1=2.0 we clearly see that the period of oscillation of the pendulum is larger than for the approximate model. For yp1=6.25 we find non-sinusoidal oscillation of y with distinctly flattened peaks (below).

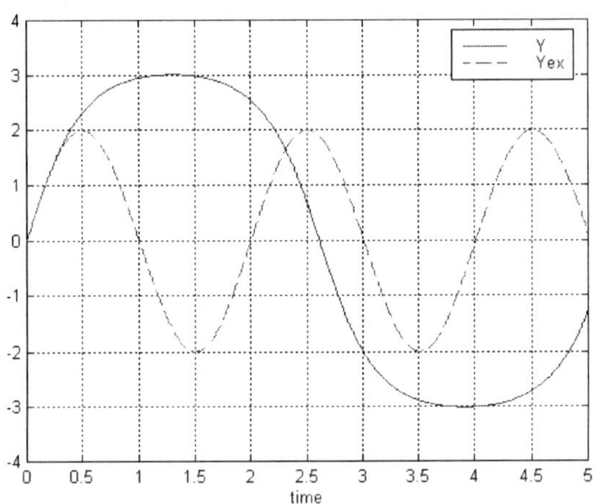

Non-Periodic Solutions

We now revert to *ex202* and focus on the behavior at larger initial velocity. Putting yp1=6.3 you will find (next figure) that the angle increases monotonically with time, although the angular velocity still varies nearly periodically. Under these conditions the pendulum no longer moves back and forth but makes full turns in the same direction forever.

In this plot, we notice that there are rather flat regions close to $y = \pi$, 3π, and so on. Of course, these particular angles would correspond to the pendulum being in a vertical position, i.e. in unstable equilibrium.

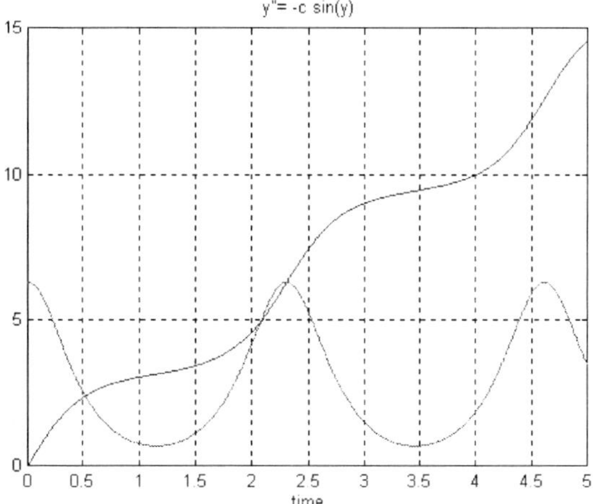

By physical arguments we can establish at what initial velocity the pendulum only just turns over (figure below). At the lowest point, the velocity is Ly_{p1} and the kinetic energy of the pendulum thus becomes $mL^2 y_{p1}^2 / 2$, while the potential energy is zero.

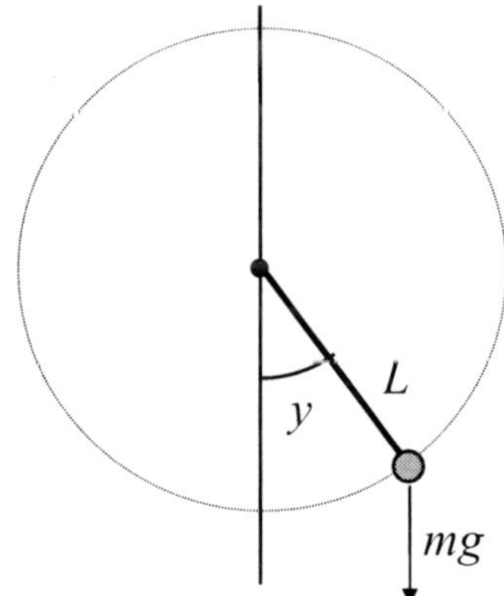

At the angle $y = \pi$ the velocity, and hence the kinetic energy, vanishes and the total energy becomes *2Lmg*. Equating the total energies we obtain

$$\frac{mL^2 y_{p1}^2}{2} = 2Lmg, \quad y_{p1} = \sqrt{4\,g/L} \cong 6.26418390534633$$

with the value of g/L adopted here. The following figure includes the solution for yp1=6.26418391, which evidently comes close to unstable equilibrium. This is clear from the constancy of the angle y and the vanishing angular velocity. We also note that the final angle is close to π, as you can verify better by displaying the matrix Y.

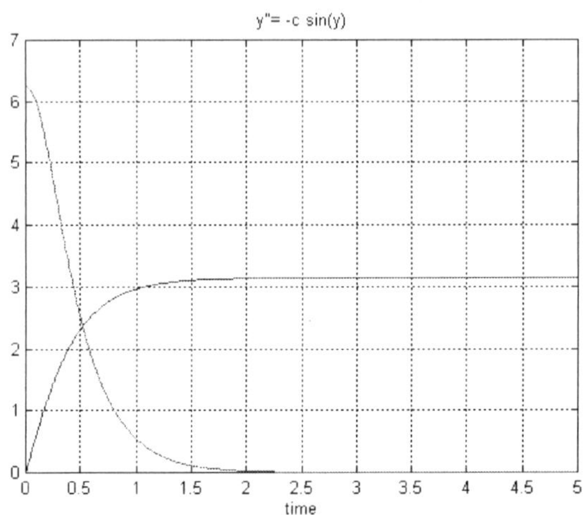

We have now solved a non-linear ODE of the second order and verified that the solution is as could be expected, both at small amplitudes and in the extreme case of unstable equilibrium. It should be clear that the MATLAB tool for ODEs is efficient and potentially useful in other problems of practical interest.

Exercises

❑ Modify *ex202a* and the function *pend* to solve the *spring* problem (*ex201*) in an alternative way. Can you detect any difference with respect to the analytic solution? Refine the test by adding a plot of Y(:,1)-Yex. Repeat for a larger amplitude value.

❑ Using *ex202* as a template, solve the second-order differential equation $y''-y(4x^2-2)=0$ by RKF with the initial conditions y1=1, yp1=0 over the region $0<x<3$. Repeat over the region $-3<x<3$ with the same initial values. Have you seen this solution before?

❑ Using *ex202* as a template, solve $y''+2y-2\sin(x)^2=0$ by RKF with the initial conditions y1=1, yp1=0 over the region $0<x<10$. Then try y1=0, yp1=1 to see a different type of solution.

❑ Change the restoring force in *ex201* to be $-c(y+0.1y^2)$. How does the period of oscillation change as the initial angular velocity increases? Make several runs with the plot zoomed around the second zero.

❑ Use *ex202* to investigate how the period length of the pendulum oscillation changes as the initial angular velocity increases. Make several runs with the plot zoomed around the second zero.

❑ After the model of *ex202*, solve the equation $y''=\sqrt{1+y'^2}$ with the initial conditions y1=0, yp1=-2 over the interval $0<x<3$. The solution shows the shape of a chain suspended between two supports (*catenary*). Use trial and error to make the support at the right end level with the one at $x=0$.

21 Symbolic Solutions to ODEs

One of the most amazing applications of symbolic methods is the solution of ordinary differential equations. By a single command we can manage both 1st- and 2nd-order equations. MATLAB exploits a large fraction of all known procedures for solving ODEs.

First-Order, Linear ODEs

In the preceding chapters on numeric methods (pp.134-157) we compared approximate and analytic solutions. We may solve the elementary equation

$$\frac{dy}{dt} + y = 0$$

analytically by typing

```
clear all,  y=dsolve( 'Dy+y=0' )                    { Dy≡dy/dt }
```

either at the MATLAB prompt or in a script file. Notice that we do not need the *syms* declaration, because the program assumes the independent variable to be *t*. Notice that we now need quotes (') to enclose the equation in its full form, including the equals sign. The resulting solution below contains an arbitrary constant *C1*, since we have not specified any initial condition.

```
1/exp(t)*C1
```

Using a similar command we may add an initial condition, either a numeric or a symbolic one. In the latter case we would type

```
clear all,  y=dsolve( 'Dy+y=0',  'y(a)= b' );  sol=simple(y);  sol
```

The first character string expresses the ODE, and the second one the initial condition. You will see at a glance that the new solution fits the initial condition.

```
b*exp(-t+a)
```

Another example that we treated numerically (p.137) may be solved symbolically by the command

clear all, y=dsolve('Dy+x^2*y=1+x^3', **'x'**); pretty(y)

where the *last* argument in parenthesis indicates that we wish to consider x (rather than the default t) to be the independent variable. This yields the following expression

```
                   3
x + exp(- 1/3 x ) C1
```

In order to obtain this solution in terms of initial values we repeat the preceding command, inserting a string as follows.

clear all, y=dsolve('Dy+x^2*y=1+x^3', **'y(x1)= y1'**, 'x'); pretty(y)

The solution

```
                   3
      exp(- 1/3 x ) (x1 - y1)
x - -----------------------
                          3
           exp(- 1/3 x1 )
```

does not seem to be identical to the one we used on p.139, but if you *expand* both expressions and then apply *simple*, they become equal.

Let us now try another example of a first-order ODE, i.e.

clear all, y=dsolve('Dy+y/(x+1)=(x-1)*sin(x)', 'y(0)= 1', 'x'); pretty(y)

which yields the answer

```
     2
    x  cos(x) - 3 cos(x) - 2 sin(x) x + 2
  - -------------------------------------
                     x + 1
```

Since this solution does not contain an arbitrary constant, we may now plot it (p.74) by the command line

ezplot(y, [-4, 4], 1), grid on, title('Dy+y/(x+1)=(x-1)*sin(x)')

to obtain the figure below.

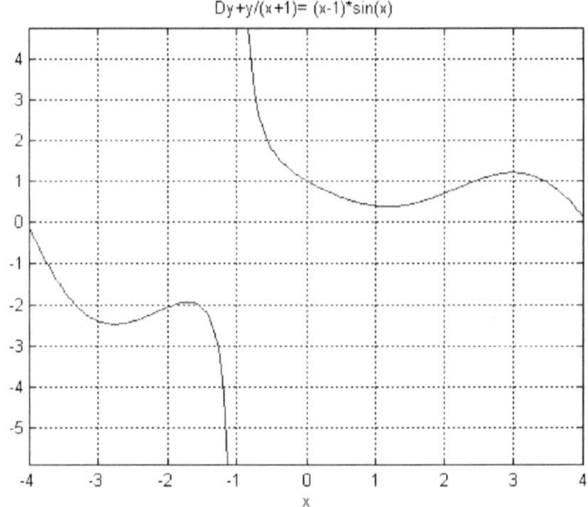

Our next example hardly appears to be more difficult to solve:

```
clear all,  y=dsolve( 'Dy+exp(x)*y= x', 'x' );  pretty(y)
```

Running this line you discover, however, that the program is unable to find a solution in terms of pre-defined functions. Instead it gives an expression comprising known functions and an integral,

```
  /
  |
  | exp(exp(x)) x dx + C1
  |
 /
-------------------------
        exp(exp(x))
```

which of course is better than nothing.

Systems of First-Order, Linear ODEs

An additional equation just becomes another argument for the *dsolve* function. As a modest example, let us take

$$\begin{cases} \dfrac{dy}{dx} - y - z = 0 \\ \dfrac{dz}{dx} - y + z = 0 \end{cases}$$

which may be coded as

```
clear all,  [y,z]=dsolve( 'Dy-y-z= 0',  'Dz-y+z= 0',  'x');
  y,  pretty(y),  z,  pretty(z)
```

resulting in a complicated solution in terms of exponentials. We may convince ourselves that the solutions are valid by typing

```
eq1=diff(y)-y-z,  eq2=diff(z)-y+z
```

To include initial conditions and to plot the solution we proceed much as before. The commands may be organized in the form of the following script file.

```
% ex211.m:  Solve a System of 1st-Order ODEs
clear all,  echo off,  hold off
[y,z]=dsolve('Dy-y-z= 0',  'Dz-y+z= 0',  'y(0)=0,  z(1)=1', 'x')
ezplot(y, [-4, 4], 1),  title('Solution for y'),  grid on
ezplot(z, [-4, 4], 2),  title('Solution for z'),  grid on
```

The last lines produce plots of the solutions, shown below for *y* (left) and *z* (right).

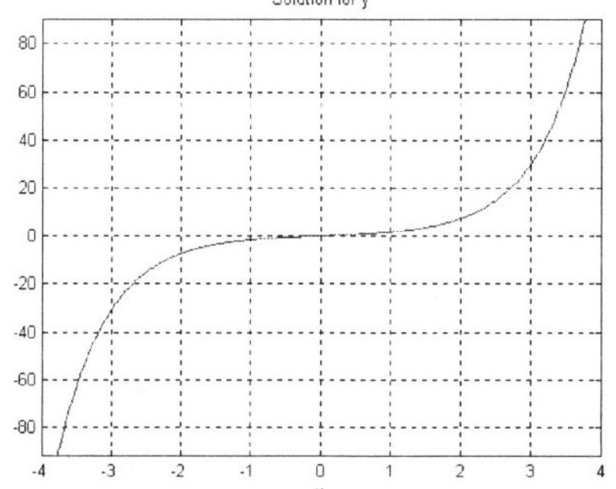

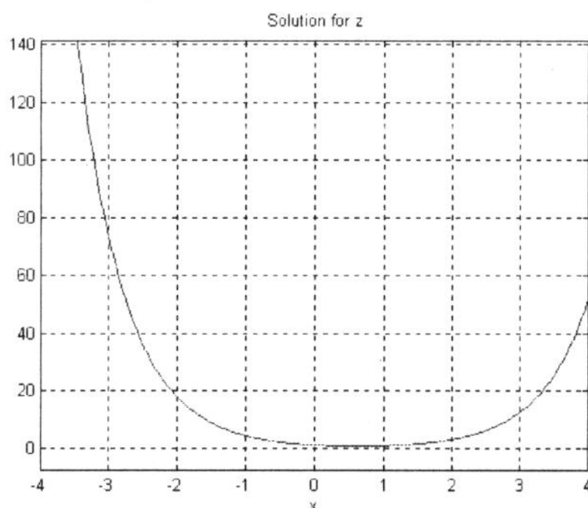

These figures confirm that the initial conditions (y(0)=0, z(1)=1) are satisfied, but we may verify this definitely by substituting values for *x* into the functions *y* and *z*. We already used *subs* on p.71 for an algebraic expression. For a solution to an ODE we would need to type

```
y0=subs(y,'x',0),  z1=subs(z,'x',1),  z1=simple(z1);  z1
```

Second-Order ODEs

A similar syntax applies in this case, with D2y for a second derivative (d^2y/dt^2). Let us try to solve the approximate equation $(\sin(y) \cong y)$ for the pendulum (p.153) as follows.

```
clear all, syms rc real
y=dsolve('D2y+rc^2*y= 0', 'y(0)= y1, Dy(0)= yp1')
```

with $rc \equiv \sqrt{c}$. We first declare rc to be real, and square it to guarantee that rc^2 will be real and non-negative. The result becomes the same as the expression we used for comparison with the numeric solution, i.e.

```
y1*cos(rc*t)+yp1/rc*sin(rc*t)
```

Let us next solve a problem from one of the exercises on p.157, namely $y''+2y-2\sin(x)^2=0$. Using the MATLAB syntax we create the file

```
% ex212.m:  Solve y"+2*y-2*sin(x)^2=0
clear all,  echo off,  hold off
y=dsolve('D2y+2*y-2*sin(x)^2=0','x')
y=simple(y); pretty(y)   % Display simplified solution y
                         % Solve and plot solution with initial conditions
clear all,  y=dsolve('D2y+2*y-2*sin(x)^2=0', 'y(0)=1, Dy(0)=0' , 'x');
ezplot(y, [0 10], 1),  grid on,  title('y(0)=1, Dy(0)=0');
clear all,  y=dsolve('D2y+2*y-2*sin(x)^2=0', 'y(0)=0, Dy(0)=1' , 'x');
ezplot(y, [0 10], 2),  grid on,  title('y(0)=0, Dy(0)=1');
```

For the general solution we obtain the expression

```
                                      1/2                 1/2
1/2 cos(2 x) + 1/2 + C1 cos(2    x) + C2 sin(2    x)
```

The first solution with initial conditions (y(0)=1, Dy(0)=0) is the simplest one, as we realize by inspecting the general expression. The second solution (in the figure below) contains contributions from the other terms and is quite different in character.

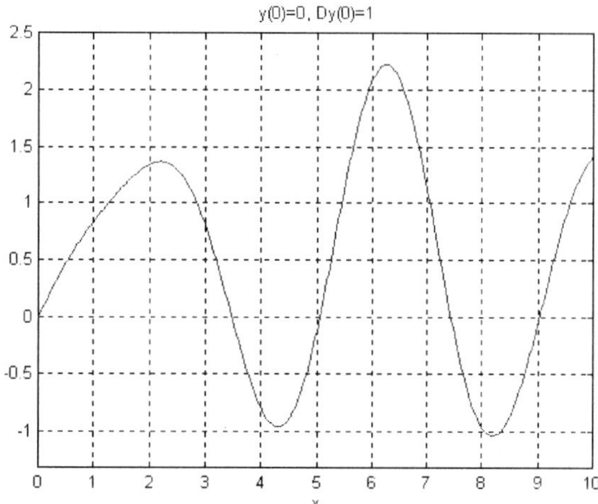

To verify the initial conditions (y(0)=0, Dy(0)=1) in the latter case it is sufficient to type

```
y0=subs(y,'x',0);  y0=simple(y0);  y0
yp0=subs( diff(y),'x',0);  yp0=simple(yp0);  yp0
```

Boundary Conditions

In the case of a linear, second-order ODE we have seen that two initial conditions, one value and one derivative, suffice to produce a unique solution. Another way to restrict the solution is to require that the curve pass through two given points. Such restrictions on a solution are known as *boundary* conditions. As an example, let us solve a simple ODE by the line

```
clear all,  y= dsolve( 'D2y+ y= 0',  'y(0)= 0,  y(1)=1',  'x')
```

Here the program promptly responds by giving a single solution. If, on the other hand, we type

```
clear all,  y= dsolve( 'D2y+y= 0',  'y(0)= 0,  y(pi)=0',  'x')
```

the program correctly concludes that there are infinitely many solutions, corresponding to the arbitrary values of the constant *C2*.

As a final example of a ODE with boundary conditions, let us explore

$y''+y'-y=\exp(x)$

by the following file.

```
% ex213.m:  Solve d2y/dx2+dy/dx-y=exp(x) with Boundary Conditions
clear all,  y=dsolve('D2y+Dy-y=exp(x)', 'y(0)= 0,  y(1)=0',  'x'),  pretty(y)
ezplot(y,[0 1], 1),  grid on,  title('d2y/dx2+dy/dx-y=exp(x)')
```

The answer in pretty form becomes

```
                          1/2             3                1/2
       (-1 + exp(1/2 5    ) exp(1/2) ) exp(1/2 (5     - 1) x)
exp(x)  - -----------------------------------------------------
                                    1/2 2
                           exp(1/2 5    )  - 1

                1/2              3               1/2                 1/2
     exp(1/2 5     ) (exp(1/2)   - exp(1/2 5     )) exp(- 1/2 (5     + 1) x)
   + ---------------------------------------------------------------------
                                        1/2 2
                               exp(1/2 5    )  - 1
```

and the following figure illustrates that the boundary conditions are in fact satisfied.

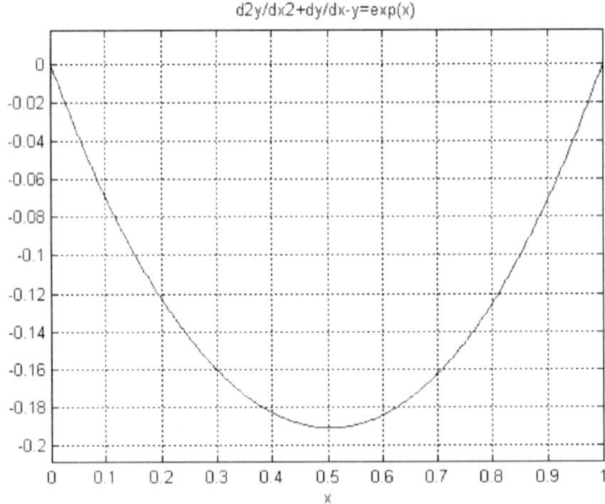

We have seen that MATLAB elegantly presents symbolic solutions to ODEs, which are easy to inspect by plots. The program only exploits known mathematical procedures that you could apply by pencil and paper, at least in principle. With such a powerful tool at our hands it is, however, not reasonable to spend months learning and applying these recipes. The program is not a “black box” in the sense that we are obliged to accept a solution on trust. We can easily verify that the solution is correct, still using MATLAB.

Exercises

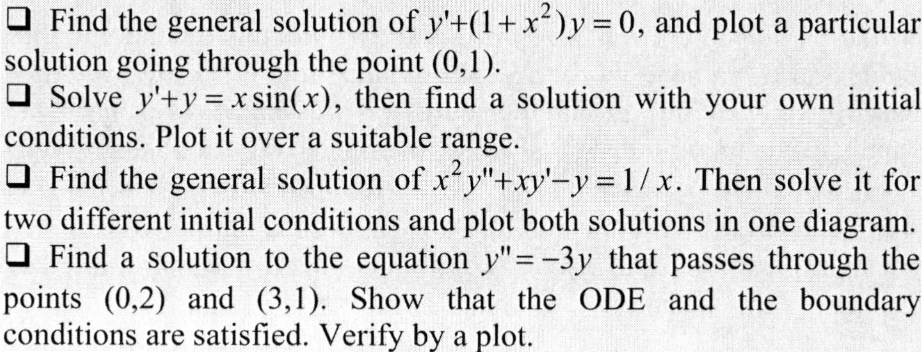

❑ Find the general solution of $y'+(1+x^2)y=0$, and plot a particular solution going through the point (0,1).
❑ Solve $y'+y=x\sin(x)$, then find a solution with your own initial conditions. Plot it over a suitable range.
❑ Find the general solution of $x^2y''+xy'-y=1/x$. Then solve it for two different initial conditions and plot both solutions in one diagram.
❑ Find a solution to the equation $y''=-3y$ that passes through the points (0,2) and (3,1). Show that the ODE and the boundary conditions are satisfied. Verify by a plot.

22 Functions of Two Variables

In the past chapters we have encountered functions of one variable, either x or t, but never both together. On the other hand, we have used functions depending on parameters, but it was understood that these finally were to be considered as constants. Here, we shall consider functions of two independent variables of the same kind, and we shall illustrate their variation by graphical means.

Surfaces

The value of a function $f(x,y)$ is determined by the values of the two variables x and y. Such a function may be displayed by a surface where each point is at a height $f(x,y)$ above the (x,y) plane. As an example, let us type a script file that displays the function

$$f(x,y) = x^2 + y^4$$

by a surface in various ways.

```
% ex221.m:  Surface Plots
clear all,  echo off,  hold off,  close all
X=-1: 2/50: 1;  Y=-1: 2/50: 1;          % Make a grid of 50x50
[X,Y]=meshgrid(X,Y);                    % Make grid matrices
F=X.^2+Y.^4;                            % Calculate function for all (x,y)
figure(1),  surf(X,Y,F),  rotate3d on,  colormap(jet)
  xlabel('x'),  ylabel('y'),  title('x^2+y^4')
figure(2),  surf(X,Y,F),  rotate3d on,  colormap(jet),  shading interp
  xlabel('x'),  ylabel('y'),  title('x^2+y^4')
figure(3),  surfl(X,Y,F),  rotate3d on,  colormap(gray),  shading interp
  xlabel('x'),  ylabel('y'),  title('x^2+y^4')
```

The first lines above define the two vectors X and Y, which specify the x- and y-values to use for calculating F. The function *meshgrid*, which we already used on p.135, transforms X and Y from vectors to square matrices. As you will see on typing X, this matrix contains all the x-values necessary for the calculations.

The first figure (below) is a three-dimensional model of the surface, where the *colormap* (from violet to red) roughly indicates the value of the function. The values occur in the order of the wavelengths of light. In this graphical representation, *meshgrid* subdivides the (x, y) domain into 50×50 cells, and the color represents an average value of the function inside each cell.

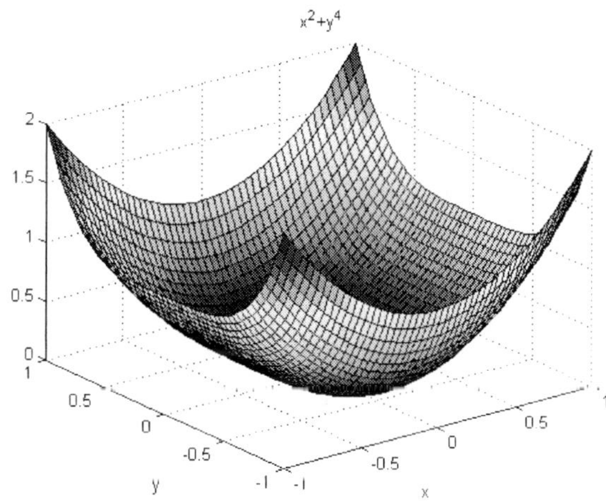

The next figure (not shown here) is a refinement of the above mosaic version, in the sense that the color values are interpolated to apply to every point, rather than to the cells. This results in a smooth variation of the color.

The third figure (below) presents the surface differently, i.e. as a diffusely scattering foil, illuminated by a lamp to enhance the perception of the shape. It is even possible to install several lamps, and of different color, as suggested by *help surfl*.

If you point at this figure and hold the left mouse button down, a dashed box will appear on the screen. Dragging the mouse you may turn this box in different orientations, and the surface follows as soon as the button is released. In this manner, we may view the surface from various angles.

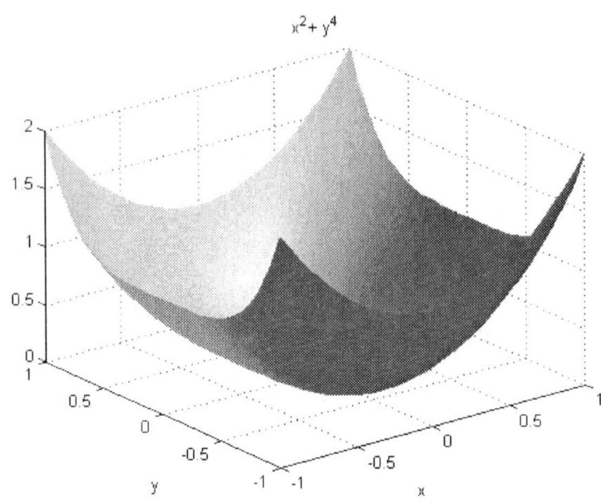

A particular viewing angle of great interest is 90° to the (x, y) plane. This option often gives the best overview of the variation of a function. The following modification of *ex221* will demonstrate two such representations.

```
% ex221a.m:  Surface and Contour Plots
...
figure(1), surf(X,Y,F), view(0,90), colormap(jet), shading interp
  xlabel('x'), ylabel('y'), title('x^2+ y^4')
figure(2), contour(X,Y,F), colormap(jet)
  xlabel('x'), ylabel('y'), title('x^2+ y^4')
```

For the first figure we request the desired direction of observation by *view*, where the first argument is the azimuthal angle to the x-axis and the second argument the elevation angle. The result is a color map (not shown here), where a given color corresponds to a specific function value. Using this map, we might draw a set of curves, each for a constant function value.

The second figure (below) consists of a set of curves corresponding to a sequence of constant values of the function. These curves are the result of data interpolation. The contour lines correspond to given values of the function, at constant increment, and the colors roughly indicate these numeric values.

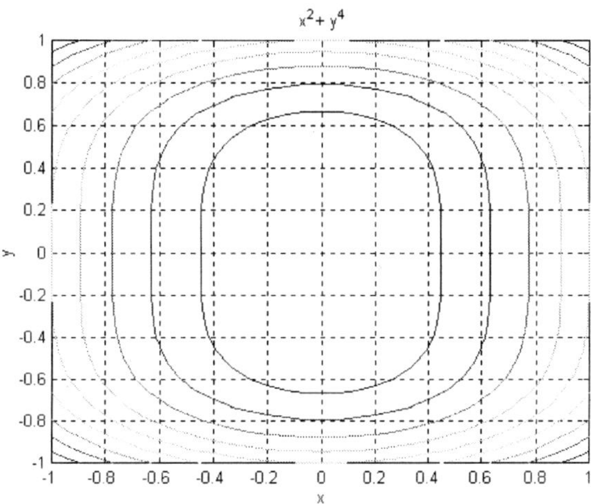

We may also display the numeric values by a device analogous to *legend* on p.49. If we modify the last two lines of *ex221a* to read

```
% ex221b.m:  Surface and Contour Plots
...
figure(2),  cs=contour(X,Y,F);  colormap(jet)
  xlabel('x'),  ylabel('y'),  title('x^2+ y^4'),  clabel(cs)
```

we obtain value labels for the contour lines, each identified by a cross.

We are also free to specify at what function *values* we want the level curves. A vector as the final argument arranges this, as follows.

```
% ex221c.m:  Surface and Contour Plots
...
figure(2), cs=contour(X,Y,F, [0:0.1:2]);  colormap(jet)
  xlabel('x'),  ylabel('y'),  title('x^2+ y^4'),  clabel(cs)
```

This leads to a large number of labels, but we may of course eliminate them by excluding clabel(cs).

It is even possible to combine two plots of the same function by the following modification of *ex221*.

```
% ex221d.m:  Combined Surface and Contour Plots
...
figure(1),  surfc(X,Y,F),  rotate3d on,  colormap(jet),  shading interp
  xlabel('x'),  ylabel('y'),  title('x^2+y^4')
```

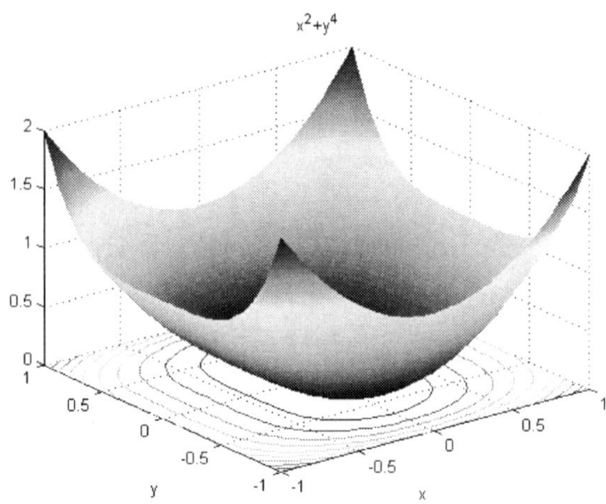

The resulting figure on the screen is colored, but the above copy shows it in a corresponding gray scale.

Zero Curves (Optional)

In the case of functions of one variable, there were *points* of zero value. Functions of two variables, on the other hand, may be zero along *lines* in the (x, y) plane.

Let us study the behavior of the function

$$f(x, y) = \exp(-x^2 - y^2) - \sin(y)^2$$

which we introduce for F in the following script file. In this case, we supply the vector [0 0] to obtain a contour for zero function value.

```
% ex222.m: Contour Corresponding to the Value Zero
clear all,  echo off,  hold off, close all
X=linspace(-4, 4, 100);  Y=linspace(-1, 1, 100);
[X,Y]=meshgrid(X,Y);                        % Make grid matrices
F=exp(-X.^2-Y.^2)- sin(Y).^2;       % Calculate function for all (x,y)
figure(1),  contour(X,Y,F, [0 0]),  zoom,  xlabel('x'),  ylabel('y')
  title('Zero Curve for exp(-x^2-y^2)- sin(y)^2')
```

The curve for zero values is shown in the figure below. The curve gives the impression of being cut off at the left and right ends, and zooming confirms this. The limited number of grid points in the plot causes this illusion, and we are in fact looking at a polygon curve.

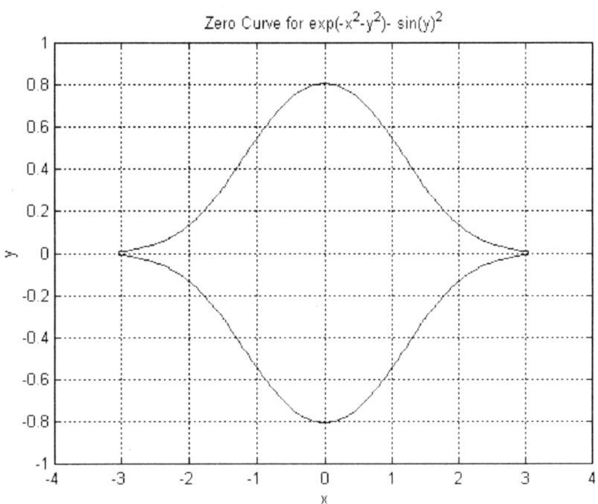

Minimum and Maximum Points

To explore minima and maxima we may first plot the function for an overview and then apply *fmins* to find the positions more accurately, as we did before (p.129). Finally, we calculate the function values at these extreme points.

Let us use the function

$$f(x, y) = \exp(-x^2 - y^2) - 1.1\exp\left(-(x-0.1)^2 - (y-0.2)^2\right)$$

for a test of these steps.

```
% ex223.m:  Surface and Contour Plots for Extreme Points
clear all,  echo off,  hold off,  close all
X=linspace(-1,1,30);  Y=X;          % Make a grid
[X,Y]=meshgrid(X,Y);                % Make grid matrices
F=exp(-X.^2-Y.^2)- 1.1*exp(-(X-0.1).^2-(Y-0.2).^2);
figure(1),  surf(X,Y,F),  view(0,90),  colormap(jet),  shading interp
  xlabel('x'),  ylabel('y')
figure(2), cs=contour(X,Y,F,20);  colormap(jet),  clabel(cs)
  xlabel('x'),  ylabel('y')
  title('exp(-x^2-y^2)- 1.1*exp(-(x-0.1)^2-(y-0.2)^2)')
```

It is clear from the contour plot below that the function has only one minimum, at roughly (0.3,0.7).

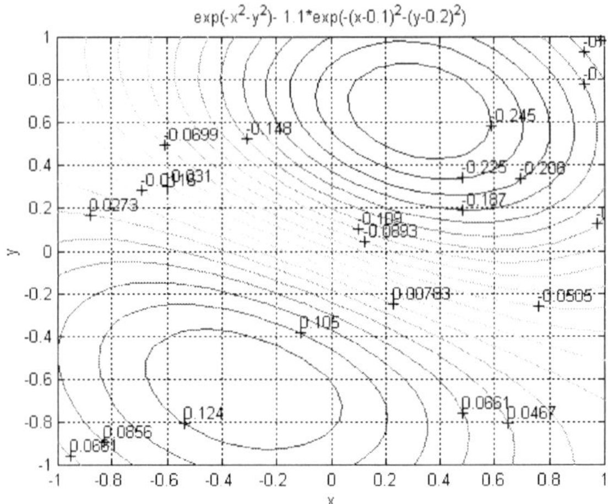

In the above script we could cite the function directly, but *fmins* refers to a special type of function file. The argument must be a vector, with one element for each variable. All the other variables in the function file are effectively scalar, but we may just as well use the vector form for the entire file. This is convenient, since we can just copy the expression from the script to this function file.

```
% expexp.m:  Function for a Minimum
function F=expexp(V);
X=V(1);  Y=V(2);
F=exp(-X.^2-Y.^2)- 1.1*exp(-(X-0.1).^2-(Y-0.2).^2);
```

Having typed the function in the above form we could proceed with the script that will calculate the coordinates V=[x1 y1] for the minimum. As we deduced from the plots, however, this function also has a *maximum*. If a function *f* has a maximum, -*f* must have a minimum, and we only have to prepare another function file with the modifying statement F=-F at the end. This redefines the function to be the negative of the previous one.

```
% expexpn.m:  Function for Maximum
function F=expexpn(V);
X=V(1);  Y=V(2);
```

```
F=exp(-X.^2-Y.^2)- 1.1*exp(-(X-0.1).^2-(Y-0.2).^2);
F=-F;
```

Now we are ready to find both the maximum and the minimum by the following combined script.

```
% ex224.m: Minimum and Maximum
clear all,  echo off
x0y0_min=fmins('expexp',[0 0],[0 1e-6])          % x0,y0 for minimum
Fmin=expexp(x0y0_min)
x1y1_max=fmins('expexpn',[0 0],[0 1e-6])         % x1,y1 for maximum
Fmax=expexp(x1y1_max)
```

Running this file we find accurate positions and function values for both maximum and minimum, which the plots indicate only very roughly.

Exercises

❑ Plot the function $f(x,y) = \cos(x^2 + y^2)$ over the domain $-2 < x < 2$, $-2 < y < 2$. View the surfaces from different angles.

❑ Plot the function $f(x,y) = \sin(x)\cos(y)$ over the domain $-5 < x < 5$, $-5 < y < 5$. View the surfaces from different angles.

❑ Plot the zero contour of $f(x,y) = x^2 + y^4 - 0.5$.

❑ Plot the function

$f(x,y) = x^2 + y^2 + 5\sin(x+y) + 0.2x$

to investigate if the function has any maxima or minima. Then use *fmins* to determine the corresponding coordinates (x,y) and the function value at that point.

23 Complex Numbers and Functions

One of the impressive features of MATLAB is the elegant way it deals with complex numbers. If we type

```
i
```

the answer becomes 0+1.0000i, which is equal to the imaginary unit. The notation j for the imaginary unit is also valid, if you should prefer it.

In any case you must remember, however, that these symbols may have been accidentally redefined. For instance, the statement

```
for i= 1:3,  y= i^2,  end
```

defines i as a loop index, and it remains a real number until you restore the imaginary value. If the symbol i does not yield the expected response, you should either start MATLAB afresh or execute the definition

```
i= sqrt(-1)
```

Functions of a Complex Variable

Now we may define a complex number *z*, use it in an expression and evaluate the latter just as if *z* were an ordinary, real number:

```
z=1+3*i;  f=z^3/(1+z)
```

Elementary transcendental functions accept a complex argument, e.g.

```
sz=sin(z)
```

as well as the exponential function and its inverse

```
ez= exp(z),  log(ez)
```

and you will find the expected result with the latter command.

MATLAB also provides special functions for complex numbers. Thus the line

```
z=1+3*i; f=z^3/(1+z), fr=real(f), fi=imag(f)
```

yields the real and imaginary parts separately, in the form

$$f = \mathrm{Re}(f) + i\,\mathrm{Im}(f)$$

An alternative is to represent a complex number by

$$f = |f| \exp(i\varphi) = |f| (\cos(\varphi) + i \sin(\varphi))$$ ●

where $|f|$ is the magnitude (absolute value or modulus) and φ the phase angle. We obtain this form easily by the functions

```
abs(f), angle(f)
```

where the angle is given in radians.

Complex Linear Equations

The commands for manipulating matrices and solving systems of linear equations remain valid even if coefficients and the solutions are complex. We now test this feature by the following line

```
A=rand(3)+i*rand(3), B= rand(3,1)+i*rand(3,1), X= A\B, A*X-B
```

The first statement fills the matrix A with random *complex* elements. The second one produces a random column vector, for the right side of the system of equations. Next, we solve this system and obtain a complex solution. With this solution, we compare the two sides of the system of equations.

Evidently, solving a system of complex equations is no more difficult than solving one with real coefficients.

Alternating Current

Systems of linear equations with complex coefficients often occur in physics. One of the simplest applications is to find currents and voltages in an alternating current circuit. If we use complex quantities throughout, Ohm's law may be written

$$U = Z\,I$$

just as in the case of direct currents. Here, Z is the *impedance* of the circuit element, which is equal to R for a pure resistance, $j\omega L$ for an inductance (L) and $1/(j\omega C)$ for a capacitance (C). The angular frequency ω is defined as $\omega = 2\pi f$.

The sum of all complex voltages over a circuit loop must be zero, since the first and last points are the same. Any voltage sources in a loop must of course also be taken into account. Hence, the loop will give rise to a complex equation, where the voltage applied may be known and the current unknown.

Let us consider the simple case of a loop consisting of a known alternating voltage U, loaded by a resistance, an inductance and a capacitance in series. Ohm's law yields for the current in this case

$$I = U / (R + j\omega L + 1 / j\omega C).$$

The current I will then be a function of frequency, and it is interesting to present its variation by graphs. Using numeric values for the impedances the script file required becomes as follows.

```
% ex231.m:  Frequency response of an RLC loop
clear all,  echo off,  hold off,  close all
R=10;                                   % Resistance (ohm)
L=1e-3;                                 % Inductance (henry)
C=1e-6;                                 % Capacitance (farad)
U=1;                                    % Input voltage (volt)
F=logspace(1,6,500);                    % Frequency vector (Hz)
Omega=2*pi*F;                           % Angular frequency
I=U./(R+ j*Omega*L+ 1./(j*Omega*C));            % Complex current I
figure(1),  semilogx( F, abs(I)),  grid on
  xlabel('Frequency'),   ylabel('|Current|')
figure(2),  semilogx( F, angle(I)),  grid on
  xlabel('Frequency'),  ylabel('Phase angle')
```

The first lines define the properties of the electric components. It should be clear from this file how one generates a *logarithmic* sequence of frequencies and then converts that to angular frequencies. The current I is calculated according to the above formula. It is important to choose graphs of the type *semilogx*, since the variation occurs over a large interval of f.

The first figure below shows how the magnitude (amplitude) of the current varies with frequency. The curve illustrates the typical increase of the current at the critical frequency (resonance).

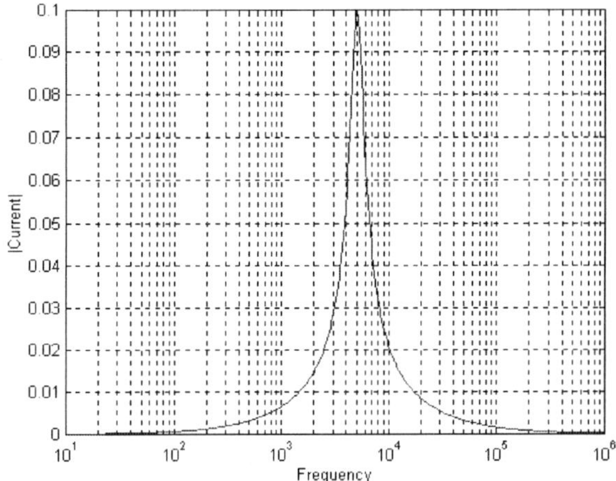

The second figure (below) demonstrates that the phase changes rapidly by the angle π at the same frequency, a behavior that is characteristic of a resonance.

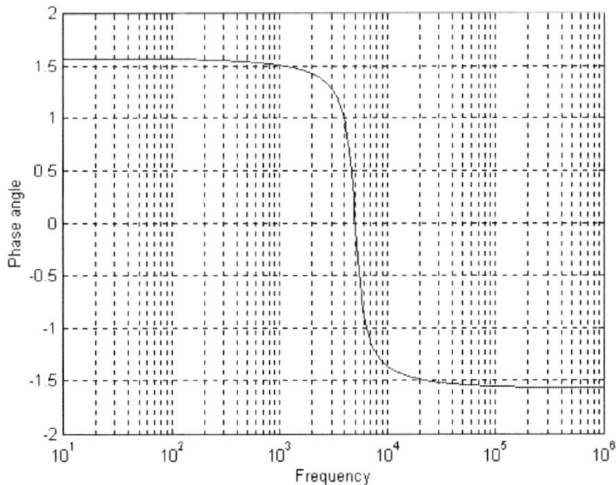

The above example concerned a circuit comprising only one loop. In a more complicated case, electrical components may be connected into a network of several closed loops. Each one of these may be treated similarly, such that each loop corresponds to a linear equation. MATLAB allows us to solve the resulting system of equations just as we did for a system with real coefficients (p.31).

Complex Zeros of a Polynomial

In an earlier chapter (p.63) we calculated some real zeros of a polynomial. A general polynomial

$$P(z) = a_0 + a_1 z + a_2 z^2 + \ldots + a_n z^n$$

may also have *complex* roots, even if the coefficients a_k are all real.

Let us consider a specific example, i.e.

$$P(z) \equiv P(x+iy) = 2.1 - 3z + 3z^2 - 3z^3 + z^4 .$$

At a zero point, both the real and the imaginary parts of $P(z)$ are equal to zero, and hence the magnitude $|P(z)|$, as we shall now illustrate by graphs.

The plot file below begins by four lines that establish a grid on the (x,y) plane. Since the 3D plots assume that the function displayed depends on X and Y, rather than on a complex Z, we next construct Z from these vectors.

The grid points usually do not coincide with the roots, and hence we could not expect to encounter values of $|P(z)|$ that are exactly zero. The magnitude Pm should rather exhibit deep minima close to the roots. In order to obtain contours as close to zero as possible we use a logarithmic scale (*logspace*) to specify the levels.

In order to inspect these points in detail we also exploit the logarithm of Pm for the second plot.

```
% ex232.m:  Plots of a Complex Polynomial
clear all,  echo off,  hold off, close all
X=linspace(-1,3,53);                          % Make a grid of nxn
Y=linspace(-2,2,61);
[X Y]=meshgrid(X,Y);                          % Make grid matrices
Z= X+i*Y;                                     % Make complex Z
Pm= abs(2.1-3*Z+ 3*Z.^2- 3*Z.^3+ Z.^4);       % Magnitude of P(z)
figure(1), contour(X,Y, Pm, logspace(-3,2,50)),  title('|P(z)|')
  xlabel('real'),  ylabel('imaginary')
```

```
figure(2), surfc(X,Y, log(Pm)), grid on, rotate3d on
  shading interp, colormap(jet), title('ln|P(z)|')
  xlabel('real'), ylabel('imaginary')
```

The plot of Pm below indicates the positions of the four roots as well as can be done, the precision being limited by the density of the (x, y) mesh.

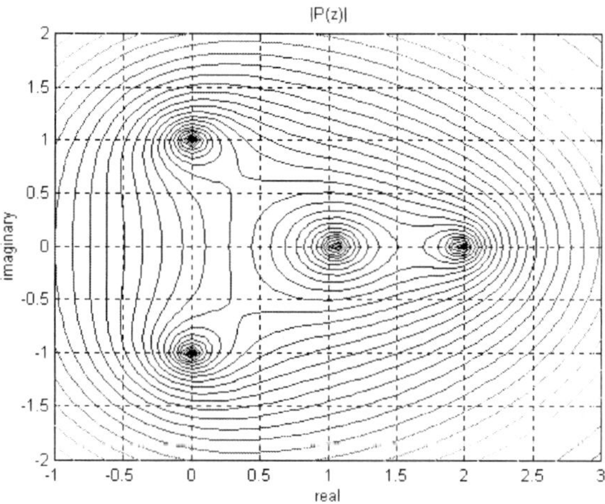

We have also included a surface plot of log(Pm), which can be rotated to display the minima in various ways.

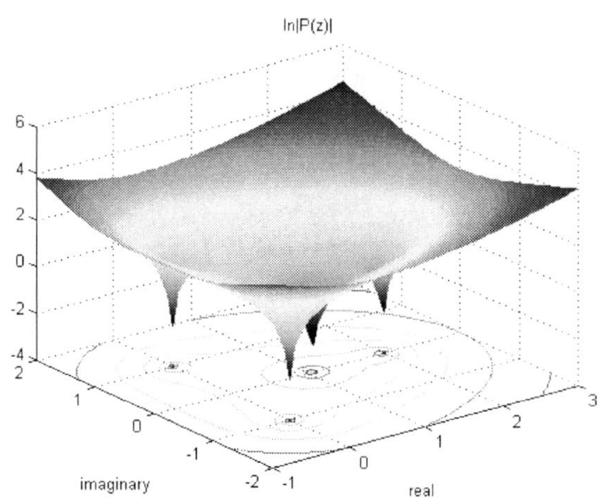

The above figure in fact shows four deep minima, pointing to the zeros indicated by the preceding plot.

Locating Complex Zeros

The above plots indicate the existence of zeros roughly at $\pm i$, 1 and 2. Since it is the mesh that limits the resolution, zooming would not improve the definition of the zeros. It is more accurate and convenient to exploit the built-in procedure for minimizing a function (pp.129, 171).

The routine *fmins* was not designed specifically for complex functions. The argument must be given as a vector V of real variables. We thus combine the two elements of V into a single complex number and use that to calculate the absolute value $|P(z)|$.

```
% polyz.m:  Magnitude of Complex Polynomial
function Pm= polyz(V);                          % Vector argument V
z= V(1)+i*V(2);
Pm= abs(2.1-3*z+ 3*z^2- 3*z^3+ z^4);    % Magnitude
```

The following script file permits us to calculate the zeros by *fmins*, using one of the estimates [0 1], [0 -1], [1 0] or [2 0] as the second argument. Each operation yields two values, which should be interpreted as the real and imaginary parts of the root.

```
% ex233.m: Calculate Zeros by Minimum of Magnitude
clear all,  echo off,  format long
z1=fmins('polyz', [0  -1], [0  1e-16])
z2=fmins('polyz', [0  1], [0  1e-16])
z3=fmins('polyz', [1  0], [0  1e-16])
z4=fmins('polyz', [2  0], [0  1e-16])
```

Here, [0 1e-16] means that we abstain (0) from viewing intermediate steps in the solution and that we request an accuracy in the root values of 1e-16.

It is interesting to verify the accuracy by investigating how $|P(z)|$ changes as you increment the real or imaginary part of a root by a small amount.

Symbolic Treatment of Complex Expressions

The symbolic algebra that we have already used (pp.69-73) also applies to complex expressions. For instance, we may type and execute each of the two lines

```
clear, syms x y, z=x+i*y;
f=z^3+3*z^2-z-3, g=z^2+4*z+3, simple(f/g)
```

The first line defines the complex z. On the second line we define the complex polynomials f and g, and finally simplify the ratio f/g. The result in terms of x and y becomes

```
x-1+i*y
```

The well-known trigonometric functions have their complex counterparts, which we verify by

```
clear, syms x y, z=x+i*y; f=sin(z); f=expand(f)
```

yielding an answer separated into its real and imaginary parts

```
sin(x)*cosh(y)+i*cos(x)*sinh(y)
```

If we now type

```
clear, syms x y real, z=x+i*y; f=exp(i*x)-cos(x)-i*sin(x); simple(f)
```

we find the answer to be zero, and hence we effectively recover the fundamental relation (p.175)

$e^{ix} = \cos(x) + i\sin(x)$.

Symbolic Algebraic Equations

We already used the *solve* procedure on p.69. The same formula applies in the case of an equation having complex roots, and even complex coefficients. As a simple example, let us attempt to solve the quadratic equation

```
clear, syms z a b, z0=solve(z^2+a*z+b)
```

Not surprisingly, the solution proves to be

```
z0 =
[ -1/2*a+1/2*(a^2-4*b)^(1/2)]
[ -1/2*a-1/2*(a^2-4*b)^(1/2)]
```

where the two roots, z0(1) and z0(2), are displayed as a column vector. We may verify these roots by typing

```
z=z0(1); eq= z^2+a*z+b, eq=simple(eq); eq
z=z0(2); eq= z^2+a*z+b, eq=simple(eq); eq
```

which yields the expected zero.

The above roots may be real or complex depending on the numerical values of a and b. For instance, the line

```
clear, syms z, z0=solve(z^2+z+1)
```

gives us two complex solutions.

Let us next apply *solve* to the fourth-degree algebraic equation that we treated numerically on p.178.

```
clear, syms z
z0=solve(2.1-3*z+ 3*z^2- 3*z^3+ z^4), vpa(z0,15)
```

Here, we transform to approximate decimal values by the function *vpa*, which we introduced on p.74. Executing these lines we quickly obtain

```
[                        1.97923260716635]
[                        1.05006588891399]
[ -.14649248040165e-1+1.00509434514067*i]
[ -.14649248040165e-1-1.00509434514067*i]
```

in excellent agreement with the earlier numeric values.

Algebraic equations up to the fourth degree can always be solved exactly in terms of algebraic functions. Only in rare cases would it be of interest to have the complete symbolic solutions, but they are quite impressive to look at. For instance, the general equation of the third degree

```
clear, syms z a b c, z0=solve(z^3+a*z^2+b*z+c), pretty(z0)
```

proves to have the following solutions. The answer is again a column vector of the roots z0(1), z0(2), and z0(3). The program reduces the display to a reasonable length by identifying two common expressions %1 and %2, which occur repeatedly in the solutions.

```
[                          1/3                                    ]
[                   1/6 %1     - 6 %2 - 1/3 a                     ]
[                                                                 ]
[          1/3                               1/2          1/3     ]
[- 1/12 %1     + 3 %2 - 1/3 a + 1/2 i 3    (1/6 %1     + 6 %2)]
[                                                                 ]
[          1/3                               1/2          1/3     ]
[- 1/12 %1     + 3 %2 - 1/3 a - 1/2 i 3    (1/6 %1     + 6 %2)]

                              3
%1 := 36 b a - 108 c - 8 a

                    3      2  2                       2         3 1/2
     + 12 (12 b    - 3 b  a   - 54 b a c + 81 c   + 12 c a )

                       2
      1/3 b - 1/9 a
%2 := --------------
            1/3
          %1
```

Exercises

- ❑ Calculate z^3 with $z = 1 + i$.
- ❑ Calculate $z \exp(z)$ with $z = 1 + i$. Also present the result in terms of magnitude and phase angle.
- ❑ Solve the complex system of equations $z_1 + 2z_2 = i$, $2z_1 + 3z_2 = 1$.
- ❑ Modify *ex231* to generate several curves for interesting values of *R*. Why not employ an outer program loop to display them all?
- ❑ Draw contours of the magnitude of $f(z) = z^3 - 3z^2 + 4z + 1$ over the domain -5<*x*<5, -2<*y*<2, using a suitable logarithmic plot. Then calculate the zeros by *fmins*.
- ❑ Solve the equation $z^2 + z + 1 = 0$ numerically after the models of *ex232* and *ex233*.
- ❑ Verify one of the solutions to *ex223*, say z1, by adding positive or negative terms, real or imaginary, to find out if $|P(z)|$ always increases.
- ❑ With a complex $z = x + i\,y$, expand $(z + 1)(z - 1)$ symbolically.

❑ Obtain the solution vector [z1 z2] for the complex system of equations $z_1 + 2z_2 = i,\ 2z_1 + 3z_2 = 1$ by symbolic means. Verify the solution, also symbolically.

❑ Try solving the equation $z^7 - z = 1$ by symbolic means.

❑ Solve $z^4 + a\,z = b$ symbolically and use *pretty* to present the solutions. Solve in particular for $a = b = 1$ and verify one of the solutions by symbolic means.

24 Fourier Series

Trigonometric functions may be represented by power series, as we have already seen (p.91). It is an astonishing fact that virtually any periodic function may be approximated by a series of sines and cosines, even if the function is discontinuous at a number of points. For a function $f(t)$ with a period of 2π we may write

$$f(t) = \frac{c_0}{2} + \sum_{n=1}^{\infty} c_n \cos(nt) + \sum_{n=1}^{\infty} s_n \sin(nt)$$ •

As explained in most textbooks on calculus we may multiply both sides by a sine or cosine factor and integrate all to obtain expressions for each of the coefficients, i.e.

$$c_n = \frac{1}{\pi} \int_{-\pi}^{+\pi} f(t) \cos(nt)\, dt$$

$$s_n = \frac{1}{\pi} \int_{-\pi}^{+\pi} f(t) \sin(nt)\, dt$$ •

It would be feasible to calculate these integrals by the Simpson method (p.95), but we shall simplify the task by using a ready-made algorithm available under MATLAB.

Continuous Functions

One of the simplest functions to expand in a Fourier series is $f(t) = \sin(t)^3$. In order to evaluate the coefficients we need to prepare one function file for $f(t)$

```
% funf.m: Function for Fourier Series Expansion
function F= funf(T);
F=sin(T).^3;
```

and one for each of the integrands:

```
% fcos.m:  Cosine Integrand
function Y=fcos(T);
global n
Y=funf(T).* cos( n*T);
```

```
% fsin.m:  Sine Integrand
function Y=fsin(T);
global n
Y=funf(T).* sin( n*T);
```

Here, we use *global* to transfer values of n to the function files, as we did before on p.122.

The final step is to integrate for the coefficients, generate the approximation Y for $f(t)$ and plot the results. The built-in integration routine *quad8* (see *help quad8*) directly gives us the coefficients c_n and s_n. The order numbers of the terms as well as the coefficients we store in the vectors N, C and S for plotting.

```
% ex241.m:  Plot of Fourier Series
clear all,  echo off,  hold off,  close all
global n
T=-pi: 0.01: pi;  Y= 0*T;          % First function for summing
for ni=0:20                         % Global n not allowed as loop index
  n=ni;
  c=quad8('fcos', -pi, pi)/pi;              % Integral for c(n)
    if n==0,  c= c/2;  end                  % First term
  s= quad8('fsin', -pi, pi)/pi;             % Integral for s(n)
  Y= Y+ c*cos(n*T)+ s*sin(n*T);             % Approximation to f(t)
  figure(1),  plot(T,Y,  T,funf(T)),  grid on
    xlabel('t'),  ylabel('funf(t)'),  pause(0.5)
  if n>0                                    % Coefficient vectors for plot
    N(n)=n;  C(n)=c;  S(n)=s;
  end
end
figure(2),  plot(N,C,'*',N,C,  N,S,'o', N,S),  grid on
  xlabel('n'),  ylabel('coefficients: c=*,  s=o')
```

On running this file you see from the first plot (below) that the Fourier approximation quickly approaches, and completely overlaps, the initial function. The figure below shows the last of these plots, obtained with all the terms.

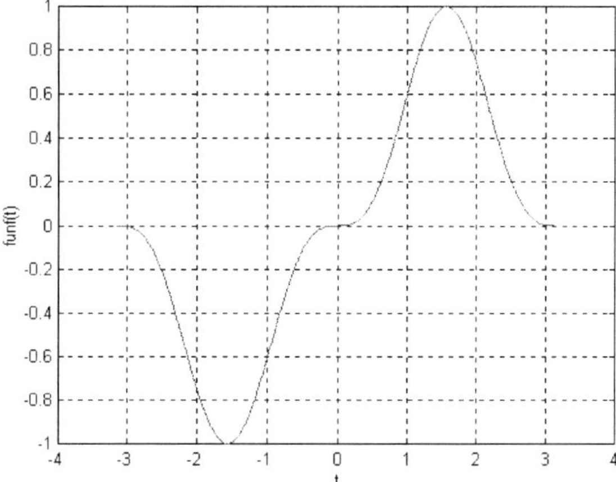

From the next plot it is evident that only the coefficients s_1 and s_3 are different from zero. This result could be expected, since the function may be expanded by elementary means as follows

$$f(t) = \sin(t)^3 \equiv (3/4)\sin(t) - (1/4)\sin(3t)$$

In this particular case, the Fourier series thus consists of only two sine terms. We may also obtain a detailed list of these coefficients by typing S after the run.

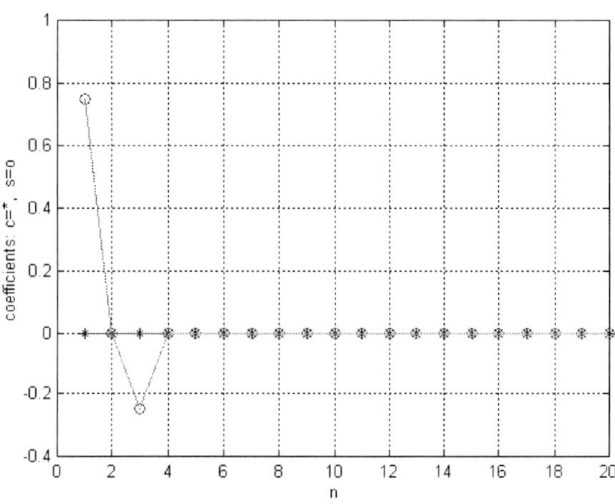

As the next example we analyze the function $f(t) = \exp(\sin(t))$. The latter is also periodic, since the sine function takes the same value at both ends of the interval. All files remain unchanged, except *funf*, which we modify by putting the comment symbol % in front of the previous function definition and by typing

```
F= exp( sin(T));
```

on the next line.

The following plot shows one period of this function, superimposed on the Fourier approximation for 20 terms, which evidently becomes indistinguishable.

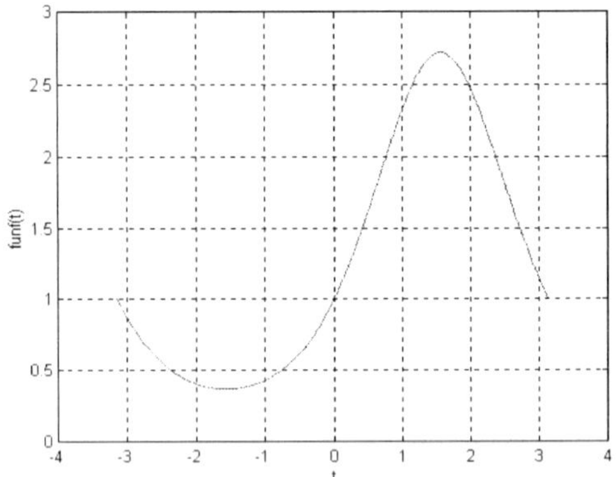

The next figure below demonstrates the high rate at which the coefficients vanish for terms of increasing order.

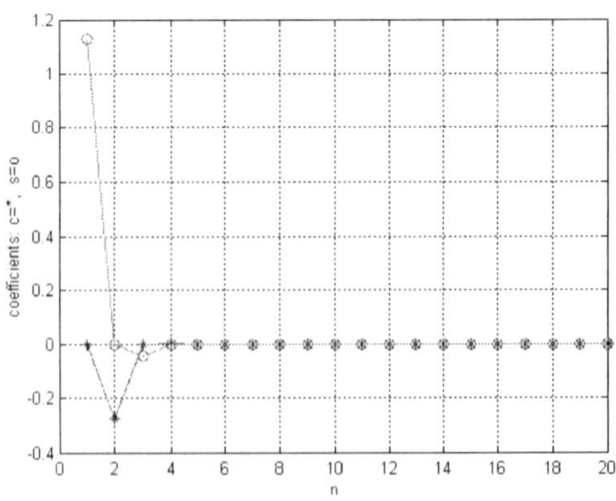

If we type the line

```
format short e,  C,  S
```

after plotting we notice that both cosine (*) and sine (o) terms are required for this expansion. Also, we observe that both eventually reach the 10^{-16} level.

Even a *polynomial* may generate a continuous, *periodic* function with a continuous derivative. We just need to choose the coefficients such that the function has the same value and the same derivative at the beginning and at the end of the period $(-\pi,\pi)$. A polynomial of the third degree satisfies these conditions if

$$f(t) = C + D\pi^2 t - Dt^3,$$

where the constants C and D are arbitrary. To illustrate this case, let us introduce the new definition

```
F= 3*pi^2*T- 3*T.^3;
```

into the file *funf*, putting % against the previous expression.

From the plots (not shown here) we see that this polynomial is a rather good approximation to a sine function, since the first sine term dominates and the series converges rapidly.

As a final example of a continuous function we choose

```
F= cos(T)+ abs( cos(T));
```

After adding this expression to *funf* and putting % against the preceding one on the list, we can now study this case.

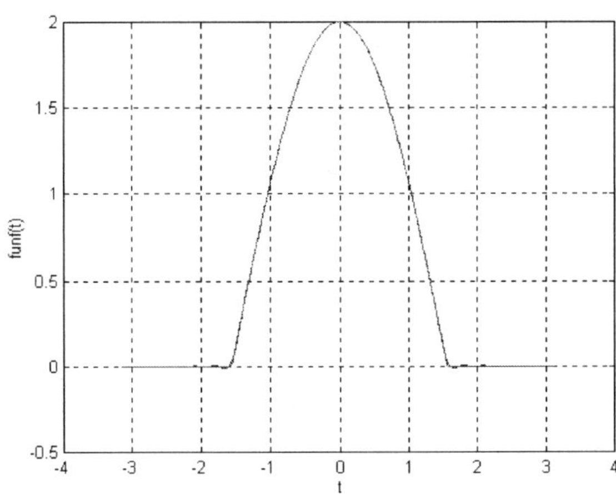

The first plot above shows the function $f(t)$ and its Fourier approximation; the difference is evidently barely detectable. Here, the function analyzed is continuous, but its derivative is discontinuous.

We see from the second figure (below) that rather many terms are involved in the expansion.

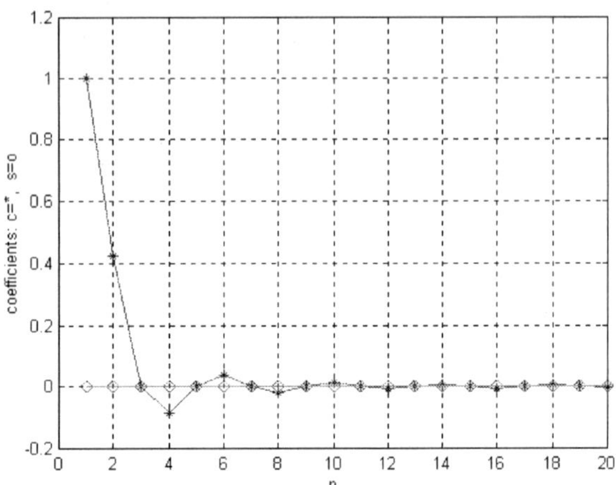

Listing the values of the coefficients we find that the elements of S are vanishingly small. This reflects the fact that $f(t)$ is now an *even* function, which makes the sine integrals exactly zero. A function $f(t)$ is called *even* if $f(-t) = f(t)$ holds for all values of t. An *odd* function, on the other hand, changes sign to make $f(-t) = -f(t)$.

Discontinuous Functions

Since Fourier expansion is not limited to continuous functions, it is fortunate that MATLAB offers a built-in discontinuous function that we may use to construct various test objects. The basic building block is the *sign* function with the following properties.

$$\text{sign}(t) = \begin{cases} -1 & \text{for } t < 0 \\ 0 & \text{for } t = 0 \\ +1 & \text{for } t > 0 \end{cases}$$

On this basis we may now define a new function, *unitstep*, which will prove useful in applications to follow.

```
% unitstep.m:  Unit Step Function
function F= unitstep(T);
F=(sign(T)+ 1) / 2;
```

Testing various input values for this function we find that unitstep(t) is zero for a negative t and unity for a positive t. Let us begin by Fourier-analyzing the function itself by typing

```
F=unitstep(T);
```

into the file *funf*, barring the previous one by the symbol %. This function is of course not periodic, but we define $f(t)$ over the interval $-\pi \le t \le \pi$, and the Fourier analysis assumes that following periods are just repetitions of the basic unit.

After running *ex241* we notice that our sum of 20 terms is not sufficient to reproduce the step of the original function (figure below). In fact, no matter how many terms we include, the under- and overshoot present in this figure will remain, although the *interval* over which it occurs will be reduced without limit.

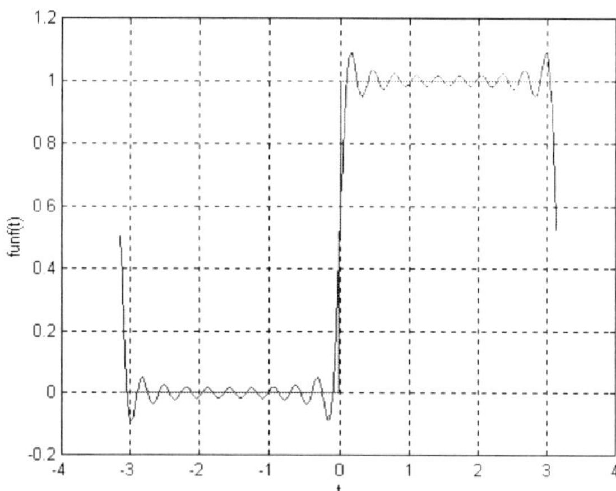

We also find that the magnitude of the coefficients (below) decreases with n much more slowly than before, which is typical of a discontinuous $f(t)$.

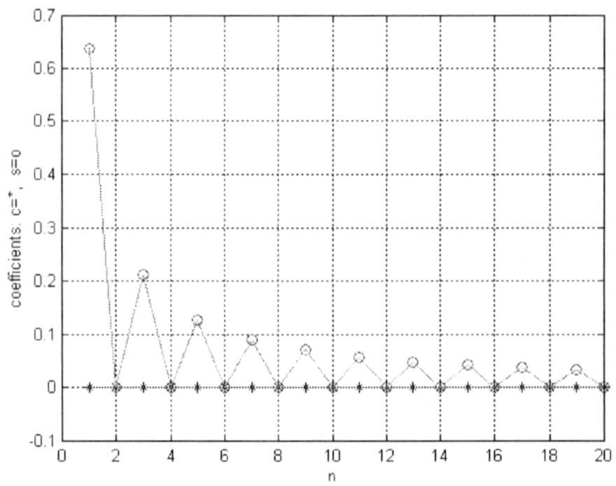

We shall now make the function less symmetric by shifting the discontinuity to the right, as follows.

```
F= unitstep(T- pi/2);
```

The main difference with respect to the previous case is that the elements of both C and S are in general non-zero, due to the lack of symmetry in the integrand.

If a continuous function is defined over a finite interval, then a *discontinuity* may arise, as in the following example.

```
F= exp(T);
```

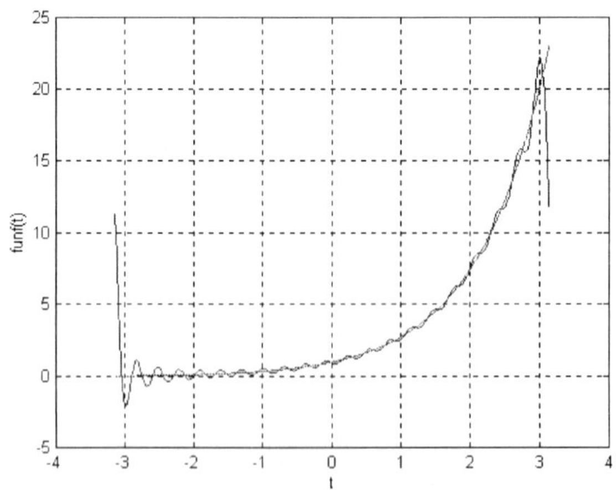

As shown above, the function values are different at the two ends of the basic interval. Much as before, the discontinuity makes the coefficients decrease slowly with n.

Arbitrary Period Length

In all of the above examples, the basic period extended from $-\pi$ to $+\pi$, but this is not a necessary restriction. We could just as well define $f(t)$ over the interval from $-L/2$ to $+L/2$. To achieve the appropriate periodic behavior it suffices to multiply t by $(2\pi/L)$ and modify the limits of integration accordingly. Thus the more general series will be

$$f(t) = c_0 + \sum_{n=1}^{\infty} c_n \cos\left(\frac{2\pi nt}{L}\right) + \sum_{n=1}^{\infty} s_n \sin\left(\frac{2\pi nt}{L}\right) \qquad \bullet$$

and the coefficients

$$c_n = \frac{2}{L}\int_{-L/2}^{+L/2} f(t)\cos\left(\frac{2\pi nt}{L}\right)dt \qquad \bullet$$

$$s_n = \frac{2}{L}\int_{-L/2}^{+L/2} f(t)\sin\left(\frac{2\pi nt}{L}\right)dt \qquad \bullet$$

Our files must now be modified to fit these expressions. Let us limit ourselves to a single example, i.e. $f(t) = \exp(-t^2)$ over the interval $-3 \le t \le 3$.

```
% funl.m: Function for Fourier Series Expansion
function F=funl(T);
F=exp(-T.^2);                                   % Gaussian with L= 6
```

```
% fcosl.m: Cosine Integrand
function F=fcosl(T);
global n L
F=funl(T).* cos(2*pi* n*T/ L);
```

```
% fsinl.m:  Sine Integrand
function F= fsinl(T);
global n L
F= funl(T).* sin(2*pi* n*T/ L);
```

The script file must now permit us to input the current period length.

```
% ex242.m:  Plot of Fourier Series
clear all,  echo off,  hold off,  close all
global n  L
L=input('Period length= ');
T=-L/2: 0.01: L/2;  Y=0*T;                % First function for summing
for ni=0:20                               % n not allowed as loop index
  n=ni;
  c=(2/L)*quad8('fcosl', -L/2, L/2);  % Integral for c(n)
    if n==0;  c= c/2;  end;
  s=(2/L)*quad8('fsinl', -L/2, L/2);   % Integral for s(n)
                                          % Fourier approximation to f(x):
  Y=Y+ c*cos(2*pi* n*T/ L)+ s*sin(2*pi* n*T/ L);
  figure(1),  plot(T,Y,'r',  T,funl(T),'b'),  xlabel('t'),  ylabel('funl'),   grid on
  pause(0.5)
  if n>0                                  % Make coefficient vectors for plot
    N(n)=n;  C(n)=c;  S(n)=s;
  end
end
figure(2),  plot(N,C,'*', N,C,  N,S,'o', N,S),  grid on
  xlabel('n'),   ylabel('c (*), s (o)')
Cm=abs(C);  Sm=abs(S);
figure(3),  semilogy(N,Cm,'*', N,Cm, N,Sm,'o', N,Sm),     grid on
  xlabel('n'),  ylabel('magnitude of coefficients: c=*,  s=o')
```

The target function *funl* in this example is the well-known Gaussian, for the interval $-3 < t < 3$, which corresponds to a period of L=6.

In the script file we have added a *logarithmic* plot, which shows the magnitude of the coefficients at large values of n. If the numeric value of a coefficient happens to be exactly zero, the corresponding point will be excluded from the plot. This is illustrated by the following figure (n=1,10).

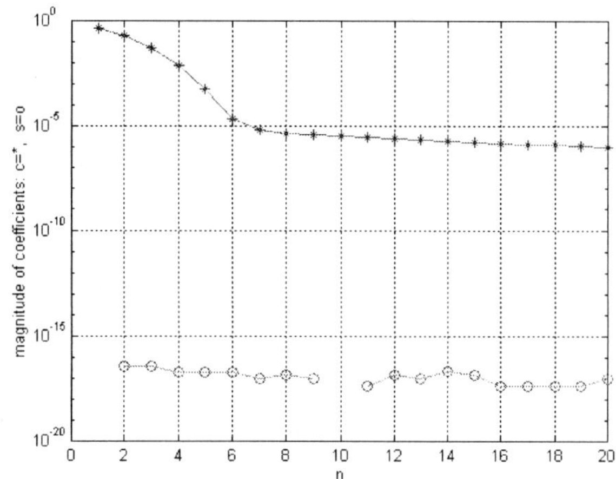

Exercises

Expand the following functions in Fourier series and analyze as far as possible the pattern of non-zero coefficients.

- $f(t) = \sin(t)\cos(t)^2$ over the interval $-\pi \le t \le \pi$.
- $f(t) = |\sin(t)|$ over the interval $-\pi \le t \le \pi$.
- $f(t) = 1 - t^2$ over the interval $-1 \le t \le 1$.
- $f(t) = \sqrt{1-t^2}$ over the interval $-1 \le t \le 1$.
- $f(t) = \begin{cases} -t & \text{for } t \le 0 \\ +t & \text{for } t > 0 \end{cases}$ over the interval $-10 \le t \le 10$.
- $f(t) = \begin{cases} t+5 & \text{for } t \le 0 \\ 0 & \text{for } t > 0 \end{cases}$ over the interval $-5 \le t \le 5$.

25 Fourier Transforms

A general Fourier series (p.193) comprises sine and cosine terms of a given period length (L) plus contributions from terms with periods $L/2$, $L/3$ and so on, in short L/n. Hence, if we add L to the independent variable t, all these terms resume their previous values. This means that the period length (period duration) of such a Fourier series is L.

It is possible to eliminate these restriction by introducing non-integral numbers for n. Mathematical textbooks on this subject prove that virtually any function of practical interest may be expressed as a Fourier sum containing terms with all possible values of n. The result is a Fourier *integral* of the form

$$f(t) = \int_0^\infty c(\nu)\cos(2\pi\nu t)d\nu + \int_0^\infty s(\nu)\sin(2\pi\nu t)d\nu$$

where ν now replaces n/L. By means of the complex exponential (p.175) this may be condensed into

$$f(t) = \int_{-\infty}^{\infty} g(\nu)\exp(i2\pi\nu t)d\nu \qquad \bullet$$

In many applications ν corresponds to the frequency of a function varying with time t.

The Fourier transform is defined in a number of different ways in the literature, and fortunately they are all equivalent. A convenient feature of the alternative chosen here is that we may express $g(\nu)$ in a symmetrical fashion, i.e.

$$g(\nu) = \int_{-\infty}^{\infty} f(t)\exp(-i2\pi\nu t)dt \qquad \bullet$$

which is easy to remember.

A Fourier transform may be used to construct a pulse $f(t)$ of limited extension in the time t. We simply express it as an integral of sines and cosines of various frequencies ν, using the complex function $g(\nu)$ to express the frequency distribution.

Conversely, we may analyze a known non-periodic function (e.g. the one below) in terms of sines and cosines. The amplitude and phase of these terms are given by the complex function $g(\nu)$.

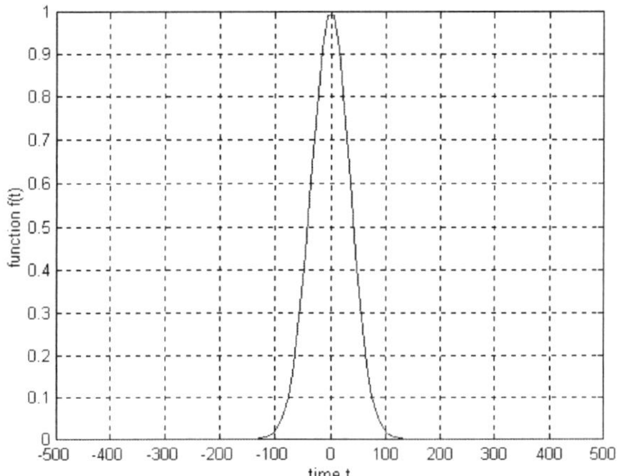

In this chapter we shall only be concerned with *real* functions $f(t)$. If we separate the complex exponential function into its trigonometric components, we realize why $g(\nu)$ still needs to be complex. Let us write

$$g(\nu) = \int_{-\infty}^{+\infty} f(t)\{\cos(-2\pi\nu t) + i\ \sin(-2\pi\nu t)\}dt =$$

$$\int_{-\infty}^{\infty} f(t)\cos(2\pi\nu t)dt - i\int_{-\infty}^{+\infty} f(t)\sin(2\pi\nu t)dt$$

If the real function $f(t)$ is *even*, i.e. $f(-t) = f(t)$, the second integrand F will be odd, i.e. $F(-t) = -F(t)$, and hence its integral becomes zero. In this case, then, $g(\nu)$ will be real.

If the real function $f(t)$ should be *odd*, however, it is the first integral that vanishes and $g(\nu)$ becomes purely imaginary. In any other situation, $g(\nu)$ will have both real and imaginary parts.

In many simple cases it is possible to find exact expressions for the transforms. Functions of the type $f(t) = p(t)\exp(-i\,2\pi\nu t)$ may, for instance, be integrated exactly with respect to t, if $p(t)$ is a polynomial. The same is true if $p(t)$ is an exponential function.

We generally assume that $f(t)$ is some sort of pulse, starting and finishing at the value zero. For instance, a pulse could be a staircase function beginning at zero, then going up and down over a number of steps and finally returning to zero.

Tables of exact Fourier transforms may be found in standard books on the subject. In many cases, however, these transforms are complicated, and often no exact expression is known.

In some situations, a pulse may even be defined by a sequence of measured data, and in that case there is no mathematical expression to be transformed. Then it is necessary to use numerical methods.

Discrete Fourier Transforms

A frequent task for the engineer is to analyze a pulse in terms of simple sinusoidal oscillations of various period lengths. Usually $f(t)$ is known at a sequence of times $t = t_0 + k\,\delta t$, where $k = 1,2,3...$ and δt is the time interval between two successive measurements. There may be thousands of measurements of this kind. This means that one has only limited knowledge of the function $f(t)$ and it would not make sense to use sophisticated methods for the evaluation of the transform integral.

A suitable mathematical representation of a measured function $f(t)$ would be a polygon through the data points. We have already performed numerical integration by means of the trapezoidal approximation (p.94). If the first and last values are zero, this method is equivalent to the Riemann staircase approximation. According to this simple strategy we replace the integral by a sum as follows.

$$g(\nu_m) = \int_{-\infty}^{\infty} f(t)\exp(-i\,2\pi\nu_m t)dt \cong \delta t \sum_{k=1}^{n} f(t_k)\exp(-i\,2\pi\nu_m t_k) \qquad \bullet$$

Each sum yields the value of $g(\nu)$ at the frequency ν_m.

Simple Fourier Transform

Let us now prepare for calculating the transform of $f(t)$ in a straightforward way by means of the above sum. The following function file is sufficient for that task.

```
% sft.m:  Simple Fourier transform
function G= sft( T, F, Nu);
dt=T(2)- T(1);                              % Time increment
nnu=length( Nu);                            % Number of elements
for k=1:nnu
  G(k)=dt*sum( F.* exp( -i*2*pi* Nu(k)* T ) );
end
```

Each turn of the loop computes a new value of the transforms. The vector G with nnu elements stores the results of the transform.

Transform of a Gaussian

The script file below uses the function *sft* to transform the Gauss error function. Here we employ a time vector T containing symmetrical values with respect to zero, since exact transforms usually take $t = 0$ to be the center of the scale.

We also create a vector Nu for the frequencies by the command *linspace*, which conveniently generates 50 uniformly spaced values from 0 to $2/a$. We have defined the horizontal scale by this vector, but we do not know the range required for the transform G. The cryptic graphics command *get(gca,'Ylim')* inputs the extreme values for us.

```
% ex251.m:  Simple Fourier Transform of a Gaussian
clear all,  echo off,  hold off,  close all
ns=101;                              % Number of samples
A=1;  a=50;                          % Constants for input function
tlim=10*a;                           % Time range
dt= 2*tlim/(ns-1);                   % Time increment
T=-tlim:dt:tlim ;                    % Sample points, input variable
F=A*exp(-T.^2/ a^2);                 % Input function
figure(1),  plot(T,F),  grid on      % Show input function
  xlabel('time t'),  ylabel('function f(t)'),  axis([-tlim tlim get(gca,'Ylim')])
Nu=linspace(0, 2/a, 50);             % Frequency vector
```

```
G=sft(T, F, Nu);                              % Simple Fourier transform
W=2*pi*Nu;                                    % Angular frequency
G_ex=A*sqrt(pi)*a*exp(-W.^2/4*a^2);                % Exact transform
figure(2), plot( Nu,real(G),'o', Nu,imag(G),'*', Nu,G_ex ), grid on
  xlabel('frequency'), ylabel('Fourier transform: real(g)=o, imag(g)=*')
  axis([min(Nu) max(Nu) get(gca,'Ylim')])
```

By this simple-minded procedure we may calculate values of the transform at any frequency, and in this case we choose an interval of Nu from zero up to a maximum of $2/a$, determined by glancing at the analytic answer. We could also have found a suitable value by trial and error.

A plot of the function $f(t)$ was shown in the beginning of the chapter (p.197). The resulting transform is worth waiting for, since the agreement with the exact transform is excellent. The plots show that the numerical transform exhibits a vanishing imaginary value (*), which is what we expect from theory (p.197). We also note that the Fourier transform of a Gaussian becomes another Gaussian, but with different parameters.

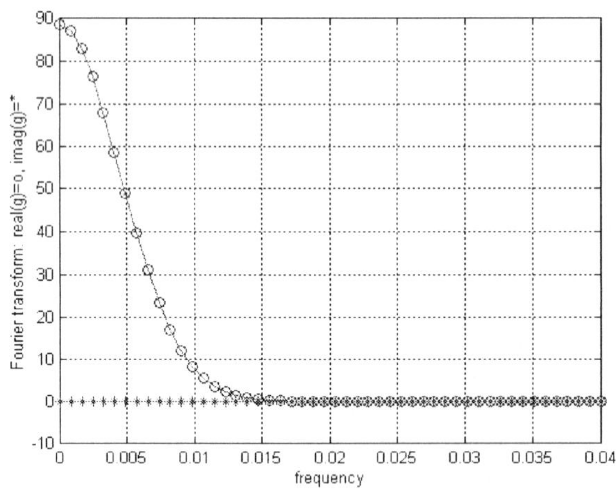

Transform of a Rectangular Pulse

For our next example we choose a function that is zero everywhere, except over the interval $-a < t < a$, where it takes the value A. For this we need the following function file. Earlier we designed the function *unitstep* (p.191), so half the work is already done.

```
% unitpulse.m:  Rectangular Pulse
function Y=unitpulse(X, x1, x2);
Y=unitstep(X- x1)- unitstep(X- x2);
```

The Fourier transform function *sft* remains the same, and we only need to change the script file *ex251* as follows (enter positions in bold). The lines to be substituted speak for themselves. We shall have several occasions to use this rectangular pulse.

```
% ex252.m: Fourier Transform of a Rectangular Pulse
clear all,  echo off,  hold off,  close all
ns=101;                                 % Number of samples
A=1;  a=1e-2;                           % Constants for function
tlim=2*a;
dt=2*tlim/(ns-1);                       % Time increment
T=-tlim:dt:tlim ;                       % Sample points, input variable
F=A*unitpulse(T, -a, a);                % Square pulse
figure(1),  plot(T,F),  grid on         % Show input function
  xlabel('time t'),  ylabel('function f(t)'),  axis([-tlim tlim get(gca,'Ylim')])
Nu=linspace(0, 5/a, 100);               % Frequency vector
G=sft(T, F, Nu);                        % Simple Fourier transform
W=2*pi*Nu;                              % Angular frequency
G_ex=2*A*sin(a*W)./W;                   % Exact transform
figure(2),  plot( Nu,real(G),'o', Nu,imag(G),'*', Nu,G_ex ),  grid on
  xlabel('frequency'), ylabel('Fourier transform: real(g)=o, imag(g)=*')
  axis([min(Nu) max(Nu) get(gca,'Ylim')])
```

Running this script file we receive an error message about division by zero, quite naturally since the first value of W vanishes. We can just ignore this, because it only means that we miss the first point on the curve for G_ex.

The figure below shows the transform of the rectangular pulse. Evidently, the answer is more complicated than in the case of the Gaussian. As before, the imaginary part (*) vanishes while the real part oscillates with diminishing amplitude toward large frequencies. The result for the real part again closely follows the full curve representing the exact transform.

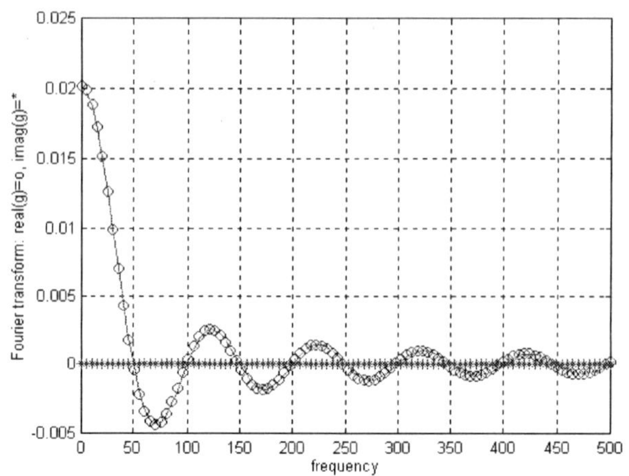

Transform of an Odd Function

In the two above examples the functions were real and *even*. A simple *odd* function, with $f(-t) = -f(t)$, is

$f(t) = A\exp(-L\,\mathrm{abs}(t))\,\mathrm{sign}(t)$

which we now substitute into a copy of *ex251* as indicated below.

```
% ex253.m:  Transform of an Odd Function
clear all,  echo off,  hold off,  close all
ns=1001;  A=1;  a=200;  tlim= 20/a;
dt=2*tlim/(ns-1);                          % Time increment
T=-tlim:dt:tlim;                           % Sample points, input variable
F=A*exp(-a* abs(T)).* sign(T);             % Odd exponential
figure(1),  plot(T,F),  grid on            % Show input function
  xlabel('time t'),  ylabel('function f(t)'),  axis([-tlim tlim get(gca,'Ylim')])
Nu=linspace(-2*a, 2*a, 100);
G=sft(T, F, Nu);                           % Simple Fourier transform
W=2*pi*Nu;                                 % Angular frequency
G_ex=-2*A*W./(a^2+W.^2);
figure(2),  plot( Nu,real(G),'o', Nu,imag(G),'*', Nu,G_ex ),  grid on
  xlabel('frequency'),  ylabel('Fourier transform: real(g)=o, imag(g)=*')
  axis([min(Nu) max(Nu) get(gca,'Ylim')])
```

The figure below shows the odd function $f(t)$ to be transformed.

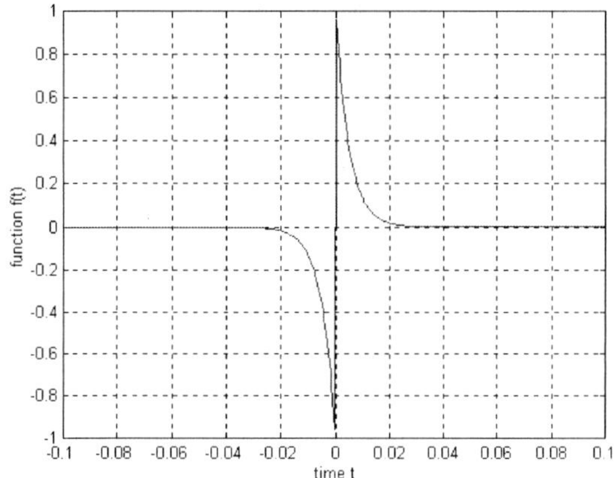

The result of the transform (below) is different from what we have seen before. The result is purely imaginary (*), as we could expect from the relation (p.197)

$$g(\nu) = \int_{\infty}^{\infty} f(t)\cos(2\pi\nu t)dt - i\int_{-\infty}^{+\infty} f(t)\sin(2\pi\nu t)dt$$

where the first integral vanishes.

In addition, we notice that the transform $g(\nu)$ has odd symmetry. This you may understand by considering the properties of the second integral, where only the sine function depends on ν.

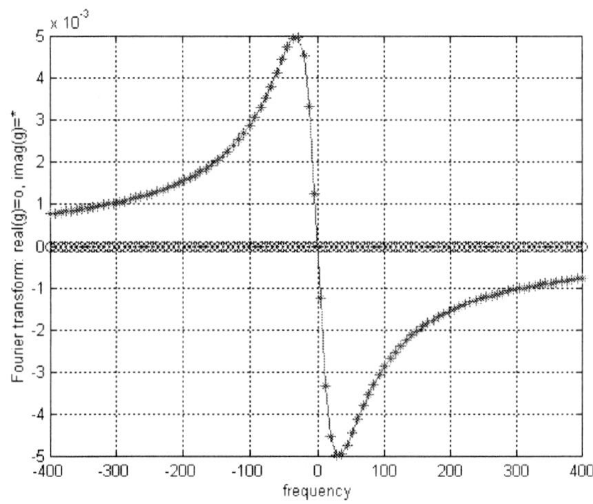

Transform of a Sine Function

As the next example, we transform a simple sine function. This corresponds to a definite frequency ν_0, so we expect to obtain an answer in the form of an infinitely narrow spike - don't we? We use a pulse consisting of exactly one period of the function, as suggested by the script below, based on *ex251*.

```
% ex254.m:  Transform of a Sine Function
clear all,  echo off,  hold off,  close all
ns=1001;  A=1;  nu0=2;  a=1/nu0/2;  w0=2*pi*nu0;
tlim=2*a;
dt=2*tlim/(ns-1);                          % Time increment
T=-tlim:dt:tlim ;                          % Sample points, input variable
F=A*sin(w0*T).*unitpulse(T, -a, a);        % One period only
figure(1),  plot(T,F),  grid on            % Show input function
  xlabel('time t'),  ylabel('function f(t)'),  axis([-tlim tlim get(gca,'Ylim')])
Nu=linspace(0, 12, 100);
G=sft(T, F, Nu);                           % Simple Fourier transform
W=2*pi*Nu;                                 % Angular frequency
G_ex=A*( sin(a*(W+w0))./(W+w0)- sin(a*(W-w0))./(W-w0));
figure(2),  plot( Nu,real(G),'o', Nu,imag(G),'*', Nu,G_ex ),  grid on
  xlabel('frequency'), ylabel('Fourier transform: real(g)=o, imag(g)=*')
  axis([min(Nu) max(Nu) get(gca,'Ylim')])
```

The function $f(x)$ to be transformed now looks as follows.

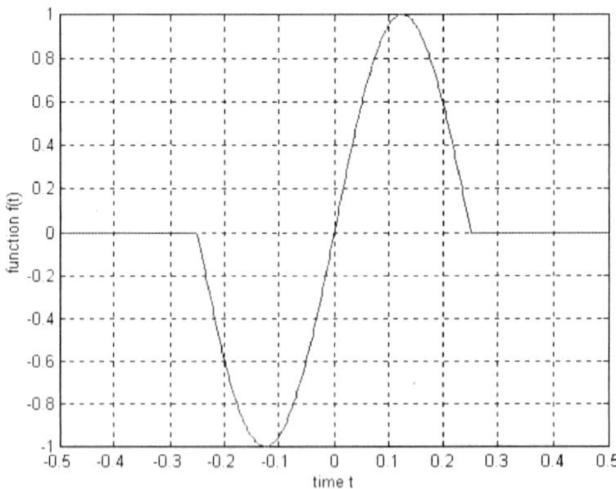

The next figure shows the Fourier transform. Instead of a well-defined frequency at 2 (Hz) we obtain a broad distribution around that

value. Also note that the transform is imaginary and that it should be an odd function of ν. The plotted points fall precisely on top of the curve representing the exact transform.

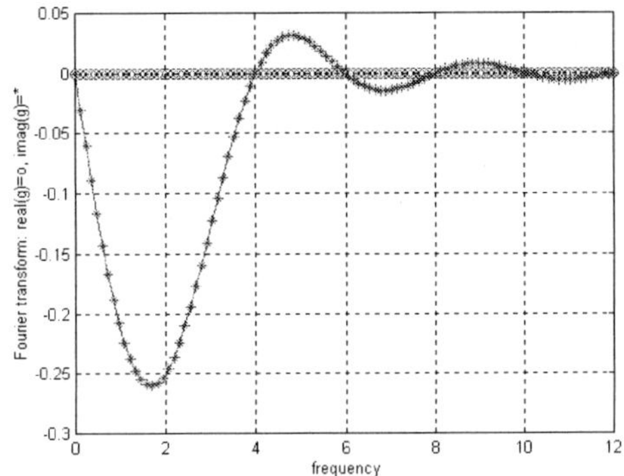

Let us now investigate what happens if we analyze a function comprising a sequence of 15 sine periods. Using *ex254* as a template we introduce the modifications below.

```
% ex254a.m: Transform of a Sine Wave Train
...
ns=1001; A=1; nu0=2; a=15/nu0/2; w0=2*pi*nu0;
...
Nu=linspace(0, 4, 200);
...
```

The function to transform consists of the wave train shown below, padded at both ends with zeros.

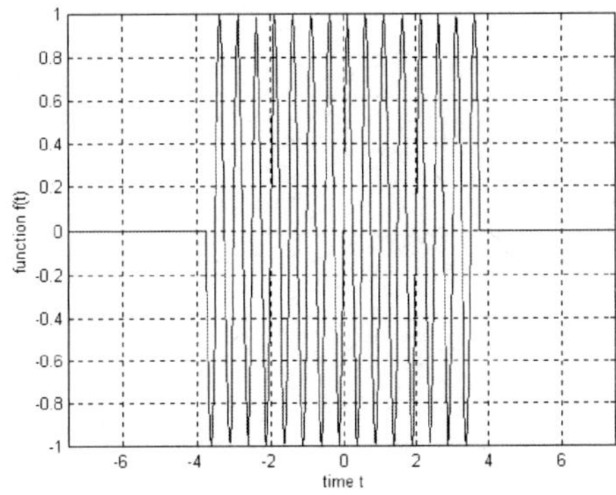

The result of this transform (next figure) agrees better with our intuitive notion. We find a negative peak at the expected frequency, and again the damped oscillation as we go away from the central frequency. These results, together with the preceding ones, suggest that a longer sequence of periods would yield an even more prominent peak, and that an infinitely long sine wave would result in an absolutely sharp frequency.

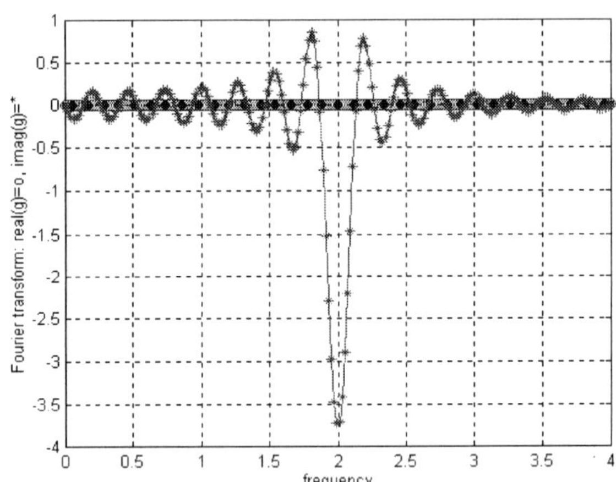

Transform of a Rectangular Wave Train

As another example of a limited pulse train we create a function alternating between two constant values. It is easy to generate this by means of the function *sign*. The file *ex254a* thus requires a few modifications and we also change from sine to cosine in order to demonstrate a difference.

```
% ex254b.m:  Transform of a Rectangular Wave Train
...
F=A*sign( cos(w0*T) ).* unitpulse(T, -a, a);
...
Nu=linspace(0, 15, 500);
...
figure(2), plot( Nu,real(G), Nu,imag(G),'--'), grid on
  xlabel('frequency'), ylabel('transform G'), legend('real', 'imaginary')
```

```
axis([min(Nu) max(Nu) get(gca,'Ylim')])
```

Even this alternating function may be transformed exactly, but we do not bother to use that rather complicated expression for comparison, and hence refrain from plotting the old *G_ex*. The result now is a real function, but a peak appears not only at the fundamental value $\nu = 2$ but also at 3, 5, and 7 times that frequency, as shown in the figure below. The additional frequencies are known as *harmonics*.

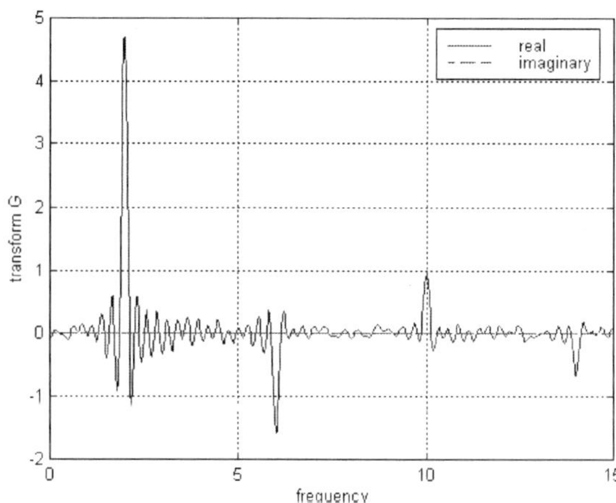

The Fourier Sum is Periodic!

The discrete Fourier transform, which we defined by the sum

$$g(\nu) = \delta t \sum_{k=1}^{n} f(t_k) \exp(-i\, 2\pi \nu t_k)$$

varies in a characteristic way with frequency. Let us assume $t_k = k\, \delta t$, which agrees with our choice of an odd number for ns. It is easy to verify this by typing

```
format short e, T
```

after running *ex253*.

For $t_1 = \delta t$, the exponent becomes $-i\, 2\pi \nu \delta t$, which yields the same function value if ν is increased by $1/\delta t$. The corresponding term is thus periodic, and the same is true of the other terms, since all exponents are multiplied by the integer k.

Let us now study this periodic behavior by changing two lines in *ex253*.

```
% ex253a.m:  Transform of an Odd Function, Periodicity
...
Nu=linspace(-50*a, 50*a, 500);  period=1/dt
...
figure(2),  plot( Nu,real(G), Nu,imag(G), Nu,G_ex,'--'),  grid on
  xlabel('frequency'),  ylabel('transform G'),  legend('real', 'imaginary')
  axis([min(Nu) max(Nu) get(gca,'Ylim')])
```

The new result for the Fourier transform is shown below. On p.203 we only viewed a narrow, central portion of the diagram. Here, we find that the transform in fact is periodic, with a period length given by $1/\delta t$, as displayed.

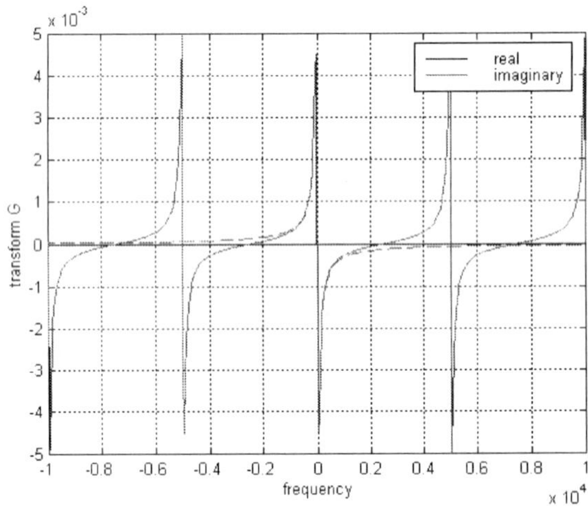

This figure also teaches us that the numeric transform agrees with the continuous (analytic) transform (dashed) only over an interval that is small with the respect to the period length. This may be understood if we realize that the continuous transform effectively exploits an infinite number of sample points (ns), yielding an infinite period length.

Analysis of a Noisy Signal

An important application of Fourier transforms is to detect a periodic signal of unknown frequency in the presence of random noise. In the chapter on fitting (p.126) we used the function *randn* to create random numbers corresponding to a given standard deviation. Let us now create a signal consisting of a cosine plus noise of much higher amplitude. This example resembles *ex254a*, now modified as follows.

```
% ex254c.m:  Transform of a Sine Wave Train with Noise
...
randn('seed', sum(100*clock));                    % Normal distribution
Noise=3*randn(1,ns);
F=A*cos(w0*T)+ Noise;
...
figure(2),  plot( Nu,real(G), Nu,imag(G),':'),  grid on
  xlabel('frequency'), ylabel('transform G'), legend('real', 'imaginary')
  axis([min(Nu) max(Nu) get(gca,'Ylim')])
```

According to the first figure below, the target function F hardly shows any signs of being periodic.

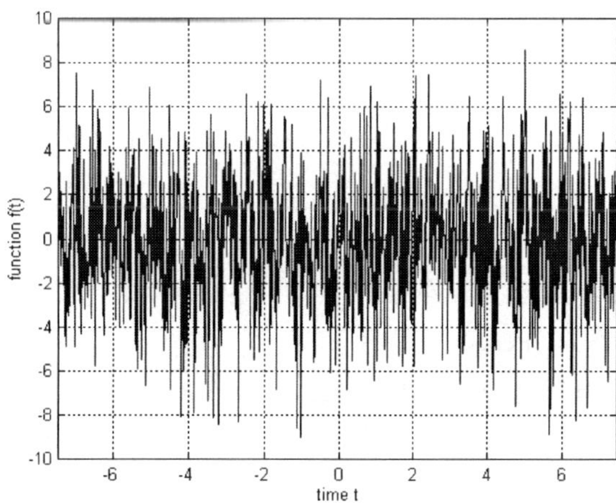

In the Fourier transform below, however, we see a distinct peak at the expected frequency, and it is *real*, as suits the transform of a real, even function.

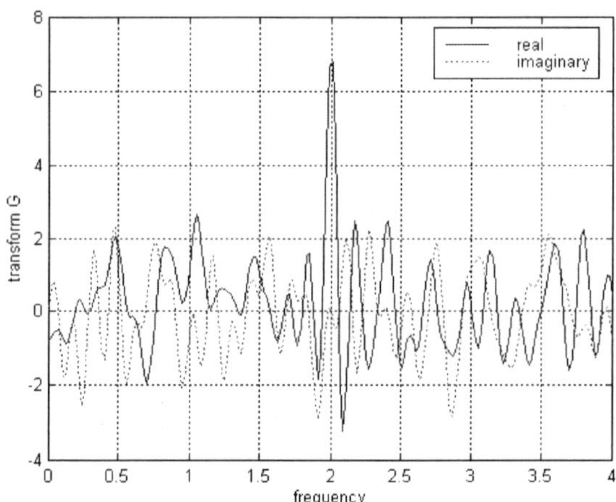

Inverse Transforms

So far we have analyzed the frequency contents of a function $f(t)$, but we may also proceed in the other direction and generate $f(t)$ from a known $g(\nu)$. The function file we need is analogous to *sft*, as is obvious from the following list.

```
% isft.m:  Inverse Fourier Transform
function F=isft( Nu, G, T);
dnu=Nu(2)- Nu(1);                    % Constant frequency increment
nt=length( T);                       % Number of time values
for k=1:nt
  F(k)=dnu*sum( G.* exp( i*2*pi* Nu* T(k) ) );
end
```

The difference with respect to *sft* is that input and output variables have been exchanged, and that the exponent now has the opposite sign. Let us combine a direct *and* an inverse transform in a single script file as follows.

```
% ex255.m:  Direct and Inverse Fourier Transforms
clear all,  echo off,  hold off,  close all
ns=101;                         % Number of samples for function
A=1;  a=50;                     % Constant for function
tlim=10*a;                      % Time range
dt=2*tlim/(ns-1);               % Time increment
```

```
T=-tlim: dt: tlim;
F=A*T.*exp(-T.^2/a^2);          % Modified Gaussian function
figure(1), plot(T,F), grid on, xlabel('t'), ylabel('target function f(t)')
Nu=linspace(-2/a, 2/a, 100);
G=sft(T, F, Nu);                % Simple Fourier transform
figure(2), plot( Nu,real(G),'o', Nu,imag(G),'*'), grid on
  xlabel('frequency'), ylabel('real(G)=o, imag(G)=*')
Ft=isft( Nu, G, T);             % Inverse transform
figure(3), plot( T,real(Ft),'o', T,imag(Ft),'*', T,F), grid on;
  xlabel('t'), ylabel('inverse transform Ft and original F (curve)')
figure(4), plot( T,real(Ft)-F,'o'), grid on
  xlabel('t'), ylabel('difference Ft-F')
```

In this script file we transform a function $f(t)$ obtained by multiplying the Gaussian by the time t (next figure).

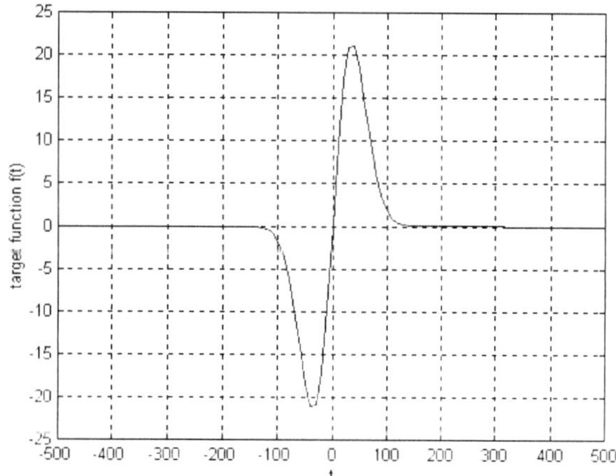

As shown below, the odd target function transforms into another function of odd symmetry. We also notice that the transform looks very similar to the original function, apart from being imaginary and having the opposite the sign.

We shall soon find out, by symbolic means, that the transform in fact happens to be of the same form as the original function. This also applied to the Gaussian itself.

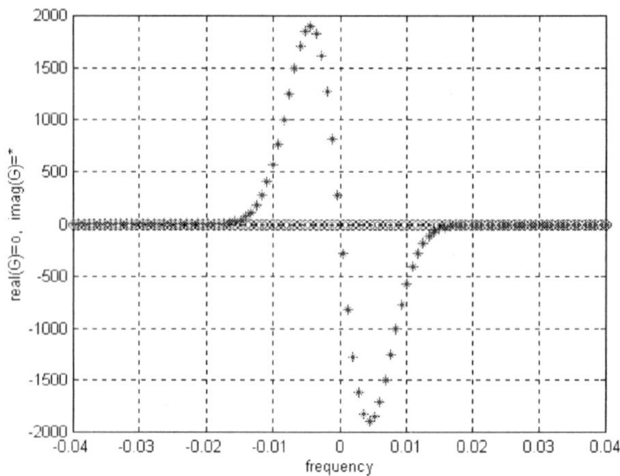

In the second part of the script file we make the inverse transform, which should recover the original function. The following plot shows that the inverse transform is real, and that it seems to agrees perfectly with the full curve for $f(t)$. In the last plot (not shown) we discover that the deviation from the original function is only about 10^{-13}.

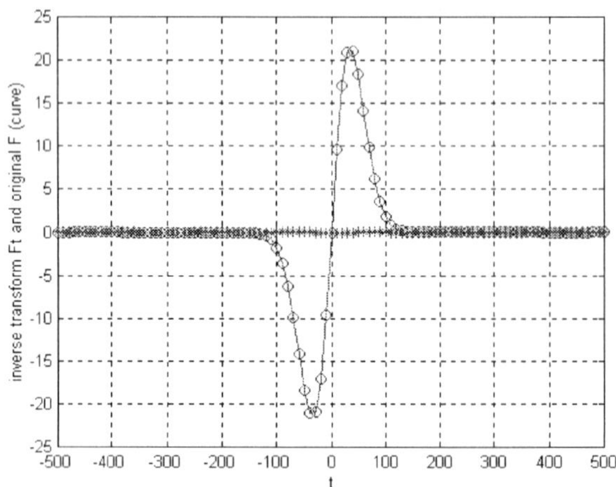

We obtain a more surprising example by extending *ex252* as follows.

```
% ex256.m: Direct and Inverse Transform of a Rectangular Pulse
clear all,  echo off,  hold off,  close all
ns=1001;                                    % Number of samples
```

```
A=1;  a=1e-2;                          % Constants for function
tlim=2*a;  dt=2*tlim/(ns-1);  T=-tlim:dt:tlim ;
F=A*unitpulse(T, -a, a);               % Square pulse
Nu=linspace(-500, 500, 200);           % Frequency vector
G=sft(T, F, Nu);                       % Simple Fourier transform
W=2*pi*Nu;                             % Angular frequency
G_ex=2*A*sin(a*W)./W;                  % Analytic transform
figure(1),  plot( Nu,real(G),'b', Nu,imag(G),'k'),  grid on
  xlabel('frequency'),  ylabel('transform: real=blue')
Ft=isft( Nu, G, T);                    % Inverse transform
figure(2),  plot( T,real(Ft),'b',  T,imag(Ft),'k',  T,F,'r'),  grid on
  xlabel('t'),  ylabel('inverse transform Ft, real=blue, original F=red')
```

When we first transformed a rectangular pulse (p.201), the result was almost indistinguishable from the curve representing the analytic expression. The figure below shows the transform valid for the present parameters, including the part corresponding to negative frequencies. The latter part is an essential ingredient, even if it is predictable in view of the expected even symmetry.

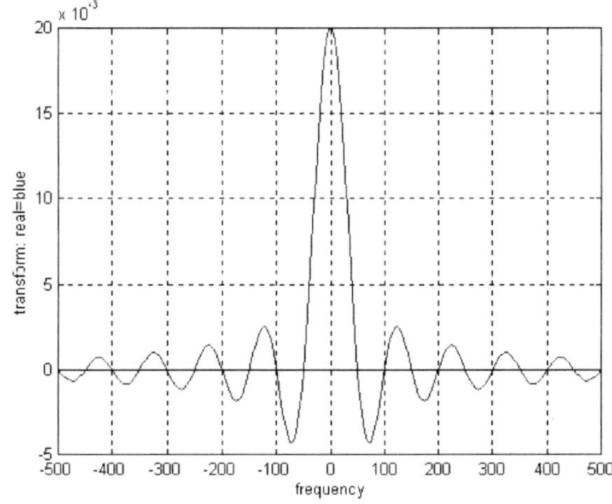

The figure below shows the inverse transform, as well as the original rectangular pulse, which is drawn in red on the screen. Evidently, the inverse transform does not agree in detail with the original, and it is reasonable to ask why this is so.

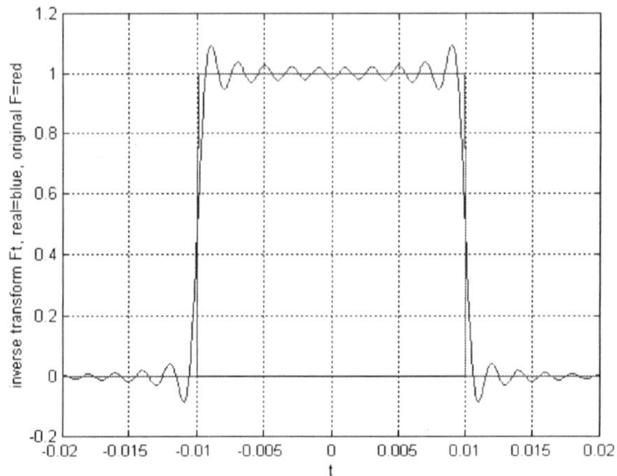

The main problem here is that we attempt to transform a function that does not vanish toward the ends of the interval chosen, but which oscillates with appreciable amplitude. Since we are forced to limit the frequency interval, we miss part of the information, i.e. that contained in the faint oscillations outside the region. We may of course increase the length of the interval, but then we must also increase the number of sample points (ns). Even if we do, the ringing overshoot at the discontinuities never vanishes, even if the peaks become narrower. This phenomenon is similar to what we saw in the case of Fourier series (p.191).

High-Speed Fourier Transforms

One may easily envisage a faster way of calculating the simple Fourier transform (p.198). It is the exponentials

$$\exp(-i\,2\pi\nu_m t_k)$$

that usually take the longest time, and there is one in every term of the sum. When the values t_k are equidistant, however, we may obtain an exponential by multiplying the preceding one by a constant factor. In fact, we need only calculate four exponential functions to carry out a transform; the remaining work is just multiplication of complex factors. Unfortunately, this procedure requires an inner as well as an outer loop in the program, a structure that MATLAB handles rather slowly. The routine *sft* may, however, be translated by a MATLAB routine into C-language, and when compiled it runs up to 30 times faster than the the version we have used above.

Fourier Transforms by Symbolic Means

According to the definition (p.196) a Fourier transform is nothing but an integral over an infinite interval. Hence, it should come as no surprise to learn that the symbolic MATLAB program can help us generate exact expressions for these.

Our first numeric example (p.199) was

$$f(t) = A\exp(-t^2/a^2)$$

and in order to find the transform we only need to type

```
syms A a t real,  g= fourier(A* exp(-t^2/a^2) )
```

where a^2 is real and positive. The answer is

```
A*(pi*a^2)^(1/2)*exp(-1/4*w^2*a^2)
```

with $w \equiv \omega = 2\pi v$.

The *inverse* operation is also at our fingertips, i.e.

```
syms A a w real,  f= ifourier(g, w, t),  f= simple(f)
```

which renders our original function.

The above results are already known, but we may also try to transform the target function in *ex255*. We only need to type

```
syms A a t real,  g= fourier(A*t* exp(-t^2/a^2) ),  g=simple(g);  g
```

to learn that the transform also is very similar to the original.

Exercises

For the following plots, use full or dashed curves instead of point symbols where suitable.

❑ Change *ex251* to make the transform include negative ν from -2/a. Does the curve become odd or even? Can you understand why by inspecting the integral expression?

❑ Modify *ex252* by putting ns=501 and making the frequency extend from -5/a to 5/a. Transform F= exp(-500*T).*unitstep(T). Why do you obtain both real and imaginary transforms?

❑ Modify *ex252* to transform a function that takes the value At/a in the interval $-a < t < a$ and zero elsewhere. Include negative ν. Compare the result to that of *ex252*.

❑ Make a copy of *ex254* and modify it to generate a transform of $\cos(\omega_0 t)$ over an interval including negative frequencies.

❑ Modify *ex254* to transform $|\sin(\omega_0 t)|$. Multiply a by 5 to include five periods of the function. Notice the positions of the peaks.

❑ Modify *ex254* by taking $f(t) = \sin(\omega_0 t) + 2\cos(3\omega_0 t)$. Calculate the transform, extended to negative frequencies. Interpret the peaks observed and notice their even or odd symmetry.

❑ Modify *ex254* again by taking $f(t) = \sin(\omega_0 t)\cos(3\omega_0 t)$ and include negative frequencies. Multiply a by 5 to include five periods of the function. Notice the positions of the peaks.

❑ Modify *ex252* by putting a=2.0 and introducing a symmetric frequency range. Transform $A\exp(-a|t|)$ and compare to the analytic transform $2Aa/\left[a^2 + \omega^2\right]$. Also try the value a=0.2.

❑ On p.204 there is an expression for the exact transform of $\sin(\omega_0 t)$, limited to the interval $-a < t < a$. Investigate by plotting this expression how the transform is influenced by the value of a. Take a to be, say, 3, 10, 30 and 100 periods.

❑ In the last section we obtained the analytic transform of the function we used in *ex255*. Include that transform in the file for comparison.

Appendix: Fast Fourier Transforms

In the preceding chapter we explored the discrete Fourier transform, which is simply a sum of products of functions. We calculated this sum in a straightforward fashion, noting that we could make it run much faster by exploiting the constant length of the intervals. The simplest recipe is, however, amply sufficient for educational purposes.

MATLAB also has built-in procedures for Fourier transforms, i.e. *fft* (*fast fourier transform*) and *ifft* (*inverse fast fourier transform).* These perform the same operations, but the calculations are better organized, and the increase in speed becomes dramatic if the sample vector comprises tens of thousands elements.

The principle of *fft* is that you take $t_k = k\,\delta t$ and let the number of sample values be $n_s = 2^m$, where m is an integer. This choice saves many operations and much time, but as we shall see there are also disadvantages.

Let us test the fast Fourier transform by the following script file, based on *ex251*.

```
% ex261.m:  Fast Fourier Transform of a Gaussian
clear all,  echo off,  hold off,  close all
ns=2^12,  A=1;    a=50;
tlim=500*a;  dt= 2*tlim/(ns-1);  T=-tlim:dt:tlim ;
F=A*exp(-T.^2/ a^2);                    % Target function
G=fft(F);                               % Fast Fourier transform
N=1:ns;
figure(1),  plot( N,real(G), N,imag(G), ':'),  grid on,  zoom on
  xlabel('index n'),  ylabel('real(g)=full curve, imag(g)=...')
figure(2),  plot( N,real(G), N,imag(G), ':', N,abs(G), '--'),  grid on
  xlabel('index n'),  ylabel('real(g)=full,  imag(g)=...,  abs(g)=---')
  zoom on,  axis([1 50 get(gca,'Ylim')])
```

The first figure (below) that results from running this file is not very informative. Most of the detail evidently is concentrated at the ends of the frequency interval, rather than at the center. We may learn more by zooming on the ends of the curve. It is clear from these enlarged plots that the real part of the transform alternates between positive and negative values, while the imaginary part oscillates much less, even if the amplitude increases as we move away from the endpoint.

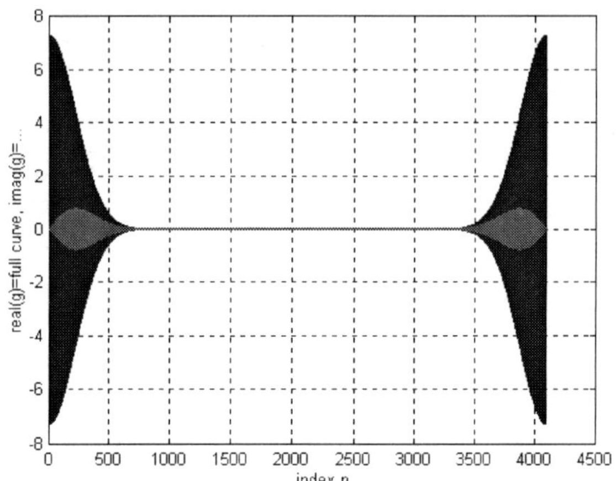

The magnitude of G, however, exhibits a certain similarity to what we obtained by direct summation (p.200). If we follow the envelope of the rapid oscillations we find that the real part takes a maximum of about 7.3. The difference between this value and the maximum we obtained by *ex251* may easily be explained. As we see from the above script file, *fft* assumes only one argument, F. This means that the time vector T is not taken into account, and strictly speaking *fft* gives us a reduced function, g/dt. The above function values should thus be multiplied by 12.21 to be comparable to those obtained before.

The second plot (below) shows the extreme left part of the preceding one. Evidently, the real part changes sign as the index increases by one unit. This in turn means that the phase angle increases by about π per index unit.

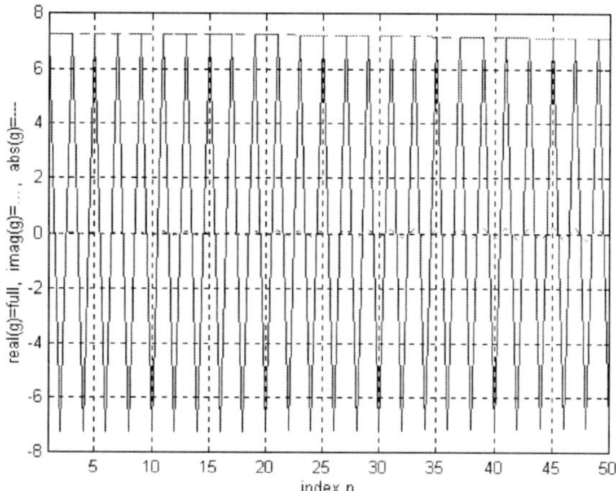

Another consequence of *fft* ignoring the time vector T is that it cannot decide whether F is even or odd. In view of this fact it is not surprising that the Gaussian does not yield a purely real and positive transform. Hence, there is no point in trying to compare the results of *fft* to those of the continuous, analytic transform.

Another shortcoming is that *fft* does not output a frequency vector Nu, but assumes that we wish to distribute the frequencies uniformly over the whole interval available (one period). Part of the speed is hence offset by unnecessary calculations.

Modified Fast Fourier Transform

There is a way (hitherto unpublished) of modifying *fft* to fit the strategy of the simple, discrete transform. The following function file implements this idea, which makes fast Fourier transforms much more useful. It takes the time and function vectors as the input and yields vectors for the frequency and the transformed function as the output.

```
% mfft.m:  Modified Fast Fourier Transform
function [Nu, G]=mfft( T, F)
ns=length(F);                          % Number of samples
N=1: ns;                               % Indices
dt=(max(T)- min(T))/(ns-1);            % Time increment
nu1=-1/2/dt;   nu2= 1/2/dt;            % Symmetric Nu scale
nnu=ns;                                % Same number out
```

```
dnu=(nu2-nu1)/(nnu-1);                    % Frequency increment
Nu= (nu1: dnu: nu2)- dnu/2;               % Frequencies
G1=dt* fft(F);                            % Raw fast Fourier transform
G2=[G1(ns/2+1: ns)  G1(1: ns/2)];         % Shift vector G1 into G2
G=i* G2.* exp(-i*(N-1)*pi*(ns+1)/ns);     % Phase corrected transform
```

Here, we multiply *fft* by the factor dt to obtain an expression for the integral. Next, we compose the vector G2 by putting the right and left halves of G1 end to end. After this modification, the essential part of the transform is at the center, but the real part still oscillates. The last line corrects this by means of a complex exponential of unit magnitude, subtracting a term $\cong \pi$ per index unit in the exponent.

We may immediately test this function by substituting it for *fft* into *ex261*. In order to focus on the non-zero part of the transform we select a central frequency range by means of *axis*. We also include the continuous transform for comparison.

```
% ex262.m:  Modified Fast Fourier Transform of a Gaussian
clear all,  echo off,  hold off,  close all
ns=2^12,  A=1;    a=50;
tlim=10*a;  dt=2*tlim/(ns-1);  T=-tlim:dt:tlim ;
F=A*exp(-T.^2/ a^2);                         % Target function
[Nu, G]=mfft(T, F);                          % Modified fast Fourier transform
W=2*pi*Nu;                                   % Angular frequency
G_ex= A*sqrt(pi)*a*exp(-W.^2/4*a^2);    % Analytic G
figure(1),  plot( Nu,real(G),'o', Nu,imag(G),'*', Nu,G_ex),  grid on
  xlabel('frequency'),  ylabel('real(g)=o, imag(g)=*, Gex=full curve')
  axis([-0.02 0.02 get(gca,'Ylim')])
```

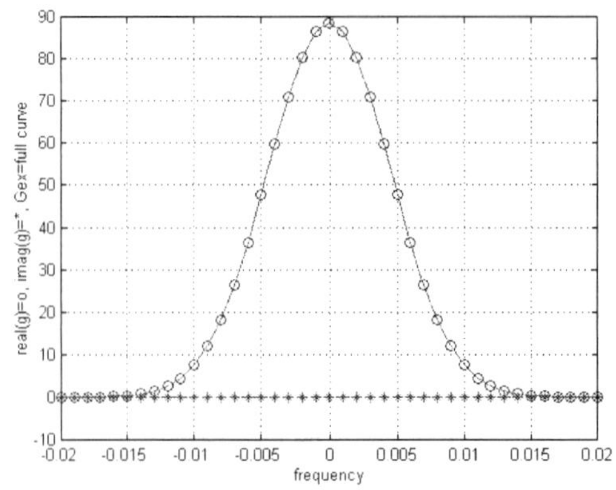

Evidently, the agreement with the continuous transform is now as good as before. Even with this modification, we are free to exploit the advantages of *fft* by using a large number of samples, ns, say 2^{14}. The Student Version 5.3 (US) does not restrict the vector length, which means that ns will only be limited by the size of the primary memory.

Exercises

❑ Modify *ex262* to transform F=A*unitpulse(T, -a, a) by the modified *fft* routine. Experiment to obtain suitable values of tlim and the frequency limit.

❑ Modify *ex262* to transform F=A*sin(w0*T).*unitpulse(T, -a, a) by the modified *fft* routine, using suitable parameters.

Appendix: Eigenvalues and Eigenvectors

If it is not your ambition to major in physics or mechanical engineering, then you may well skip this appendix, or return to it when needed. This chapter begins by comments on two particular types of systems of linear equations. One of these is related to systems of 2nd-order ODEs and is important to some applications in physics and engineering. It is also relevant to analytic geometry.

Curious Systems of Linear Equations

Normally, a system of n linear equations with n unknowns has exactly one solution. In rare situations, however, there is no solution at all. For example, the system

$$\begin{cases} 2x + 3y = 1 \\ 2x + 3y = 2 \end{cases}$$

obviously has no solution, since the left sides are equal but the right sides unequal.

In the case of a system of many equations, it may be impossible to see at a glance whether a system has a solution or not, but MATLAB decides for us. If we try to solve the following system, particularly chosen for this demonstration,

```
A=[1 2 3; 4 5 6; 7 8 9], B=[1 2 3]', X=A\B
```

we receive a warning that the matrix is close to singular. This is an extremely rare coincidence, as you may be aware after solving random systems on p.32. Unless you are born unlucky, you would never be faced with such an event.

We would have been warned about this coincidence if we had checked in advance by the command

```
determinant=det(A)
```

As you see, the value of this *determinant* is exactly zero. Such a matrix is known as *singular*.

The determinant of a matrix is in fact a particular polynomial of its elements, but its definition need not concern us now. It is enough to be able to compute it, in order to decide to what extent a solution is possible.

Even if the above example is an unlikely incident in practical life, it might be interesting to see what it implies. We can explore this special system further by the *symbolic* program. Typing the line

```
syms x y z,  [x, y, z]=solve(x+2*y+3*z-1, 4*x+5*y+6*z-2, 7*x+8*y+9*z-3)
```

gives us the result

```
x =z-1/3
y =-2*z+2/3
z =z
```

Hence, the three unknowns are not fully determined: the variable z may take any value. In other words, this system has an infinite number of solutions, involving an arbitrary parameter.

We could check this by the numeric calculation

```
z=rand,  X=[z-1/3; -2*z+2/3; z],  A*X
```

When we execute this line repeatedly, we find different values for z, but the column vector given by A*X is always equal to the original vector B. This means that the previous matrix solution, although correct for z=0, is only one of infinitely many equivalent results.

We shall now consider a special class of systems, known as *homogeneous*, where the constant terms are zero. An example of such a system is

$$\begin{cases} x+2y+3z=0 \\ 4x+5y+6z=0 \\ 7x+8y+9z=0 \end{cases}$$

Such a system has a solution only if the matrix is singular, or in the MATLAB formalism det(A)=0. We already know this to be the case with the present system. Let us now solve it by the symbolic program.

```
syms x y z,  [x, y, z]=solve(x+2*y+3*z, 4*x+5*y+6*z, 7*x+8*y+9*z)
```

This time we find the answer

```
x =z
y =-2*z
z =z
```

Again, there is an infinite number of solutions. We may verify the different solutions as before by the following line.

```
z=rand, X=[z; -2*z; z], A*X
```

Repeated runs yield a vector of zeros, with a precision of about 1e-15. Hence, we are confident to deal with cases of singular matrices, as well as the normal systems treated earlier in this book.

Special Matrix Conditions

Problems in physics and engineering often lead to a relation of the type

$$\mathbf{AX} = \lambda \mathbf{X}$$ ●

where λ is a constant. This means that the result of multiplying the column vector **X** by the matrix **A** has the effect of simply multiplying by a constant.

If we recast the above matrix equation as

$$(\mathbf{A} - \lambda \mathbf{I})\mathbf{X} = 0$$ ●

where **I** is the identity matrix (p.34), we obtain a homogeneous system of equations. This system will have solutions, provided that the determinant of the matrix $(\mathbf{A} - \lambda \mathbf{I})$ is zero. We shall see that this only occurs for a set of particular values of λ, known as *eigenvalues*.

MATLAB offers a dedicated function, *eig*, for finding the eigenvalues pertinent to a matrix. Let us test this on the matrix **A** (already defined on p.222) by running the following lines in succession. First we calculate the eigenvalues and store them in a vector E. In preparation for the test on the next line we also define the identity matrix **I**.

```
E=eig(A), I=[1 0 0; 0 1 0; 0 0 1]
```

The result of this line appears as

```
E =
   16.1168
  -1.1168
   0.0000
```

Thus there are three eigenvalues, and we notice that the third one is zero, which is to be expected in view of the singular character of the matrix **A**. Let us now display the matrix $(\mathbf{A}-\lambda\,\mathbf{I})$ for the appropriate values of λ and test whether its determinant really vanishes.

```
M1=A-E(1)*I, determinant1=det(M1)
M2=A-E(2)*I, determinant2=det(M2)
M3=A-E(3)*I, determinant3=det(M3)
```

The result for the first eigenvalue is

```
M1 =
  -15.1168    2.0000     3.0000
   4.0000   -11.1168    6.0000
   7.0000     8.0000    -7.1168
determinant1 =   1.4215e-013
```

Even if the value of the determinant is not exactly zero, it is extremely small compared to the elements of the matrix.

If we substitute an eigenvalue λ into the homogeneous system of equation $(\mathbf{A}-\lambda\,\mathbf{I})\mathbf{X}=0$, we know that solutions should exist. In fact, we may also use the function *eig* to calculate the solution vector corresponding to each eigenvalue.

```
[X, E]=eig(A)              % Matrices containing solutions and eigenvalues
```

Executing this line, we obtain the results in the form of two matrices. In the first one, shown below, the solutions appear as three column vectors.

```
X=
    0.2320     0.7858     0.4082
    0.5253     0.0868    -0.8165
    0.8187    -0.6123     0.4082
```

In order to slice this matrix into its constituent vectors we could write

```
X1=X(:,1)                                  % All rows, 1st column
```

and similarly for the other two columns.

The matrix E contains the eigenvalues in the descending diagonal, instead of in the earlier column matrix. In this representation, the first value is E(1,1), and so on.

It is easy to verify that the first of these eigenvectors actually is a solution for the current eigenvalue. It is sufficient to type

```
(A-E(1,1)*I)*X1
```

Evidently, the resulting vector elements are again small compared to the matrix elements. We could continue to verify the other two solution vectors.

Note that *eig* solves a *homogeneous* system of equations, and nevertheless it only delivers *one* solution vector for each eigenvalue. We remember from the symbolic treatment that such a system would have an *infinite* number of solutions, expressed by an arbitrary parameter. MATLAB simplifies by using the free parameter to *normalize* the vectors, i.e. give them a standard magnitude of one unit. Let us verify that this is true for the first vector.

```
magnitude=sqrt( dot(X1,X1))
```

We have thus seen that the function *eig* gives us complete access to the eigenvalues and eigenvectors related to a matrix. The tools for practical applications are hence at hand.

Balls Connected by a Spring

Previously we found that a 2nd-order ODE possesses a unique solution if we specify an initial value as well as an initial derivative. There is another frequent situation in physics and engineering where the solution is assumed to be periodic in the time variable. Such problems occur in mechanics and optics as well as in atomic and nuclear physics.

As a simplest example, let us take two objects of mass m_1 and m_2, free to move without friction along the x-axis. They are connected by a spring, such that their equilibrium positions are x_{10} and x_{20}. We denote the deviations from these equilibrium positions by x_1 and x_2 and assume that any displacement from equilibrium produces a force

$k(x_2 - x_1)$ on the first object, and an equally strong force of opposite sign on the second one.

According to Newton's law of motion we have the following ODEs for the two objects.

$$\begin{cases} m_1 \dfrac{d^2 x_1}{dt^2} = k(x_2 - x_1) \\ m_2 \dfrac{d^2 x_2}{dt^2} = k(x_1 - x_2) \end{cases}$$

Guided by the analogy with the particle on a spring (p.148) we attempt solutions of the form

$x_1 = c_1 \cos(\omega t)$ and $x_2 = c_2 \cos(\omega t)$.

Substituting these functions into the above ODEs, then dividing them by m_1 and m_2 respectively we obtain

$$\begin{cases} (k/m_1)c_1 - (k/m_1)c_2 = c_1\omega^2 \\ -(k/m_2)c_1 + (k/m_2)c_2 = c_2\omega^2 \end{cases}$$

The above system may be expressed in matrix form as follows

$$\begin{bmatrix} k/m_1 & -k/m_1 \\ -k/m_2 & k/m_2 \end{bmatrix} \begin{bmatrix} c_1 \\ c_2 \end{bmatrix} = \omega^2 \begin{bmatrix} c_1 \\ c_2 \end{bmatrix}$$

or in short

$$\mathbf{AC} = \lambda \mathbf{C}$$

This type of equation occurs again and again in the analysis of vibrating systems of the most varying kinds. Typically, **A** contains the properties of the system and λ is a function of angular frequency.

Symbolic Treatment of Eigenvalues

As we have seen earlier in this chapter, $\mathbf{AC} = \lambda \mathbf{C}$ is equivalent to a homogeneous system of equations, which may be solved by the numeric MATLAB function *eig*. In order to exploit that tool, however, we would need to assign numeric values to the parameters k, m_1, and m_2. The symbolic program also contains the corresponding eigenvalue function, however, as we shall now demonstrate by a single line.

```
syms k m1 m2,  A=[k/m1 -k/m1; -k/m2 k/m2],  [C,E]=eig(A)
```

From the result for C below we immediately notice a difference between the numeric and the symbolic *eig* function. In the symbolic version, the vectors are not normalized to unit magnitude. We should keep in mind, however, that if you multiply a given eigenvector C by an arbitrary constant, it still constitutes a solution.

```
C =
[ -1/m1*m2,          1]
[        1,          1]

E =
[ k*(m1+m2)/m1/m2,              0]
[               0,              0]
```

(The program occasionally reports an equivalent solution, where the first and second columns are interchanged.) The first eigenvector is associated with a non-zero eigenvalue, which gives us

$$\mathbf{E}(1,1) = \omega^2 = k\,\frac{m_1 + m_2}{m_1\, m_2}$$

The corresponding matrix of eigenvectors becomes, after multiplication by m_1,

```
m1*C=
[ -m2,   m1]
[  m1,   m1]
```

The first column vector implies that $c_1 = -m_2$ and $c_2 = m_1$. This indicates that the balls move in opposite directions, in such a way that the total momentum remains zero.

The second solution is clearly unrealistic. It implies that $c_1 = c_2$, which means that the balls would move in the same direction, in conflict with the principle of conservation of momentum. The corresponding eigenvalue definitely expresses that this mode of periodic motion is excluded, since $\mathbf{E}(2,2) = \omega^2 = 0$.

Large Spring-Coupled Systems

Having studied the simple example involving two balls we can imagine what form the equations will take for a larger system of balls connected by springs. We need not derive the equations of motion in detail but just notice that masses and spring constants will appear in the matrix elements.

We shall use random numbers to generate the matrix **A**, but to simplify the results we shall make it symmetrical. We can achieve this by taking

$$\mathbf{A} = (\mathbf{M} + \mathbf{M}')/2$$

since interchanging the indices of any element $a_{ij} = (m_{ij} + m_{ji})/2$ produces the same value. Let us thus simulate a large coupled system by the following line.

```
M=rand(7);  A=(M+M')/2;  [C,E]=eig(A)
```

Apart from the fact that some eigenvalues became negative, which would not be real-world results, we recognize the main features from the previous simple example. Using the same procedure as before, we may verify that each eigenvector satisfies the fundamental equation $\mathbf{AC} = \lambda\mathbf{C}$ for each of the eigenvalues.

Quadratic Forms

We should not leave this chapter without noting that eigenvalue methods also are useful for other important purposes. Polynomials containing quadratic terms often occur in geometry and also in physics and engineering. The equation

$$4x^2 + 6xy + 8y^2 = 1$$

or in short $Q(x, y) = 1$ defines a conic section, but we cannot see at a glance whether this curve is an ellipse or a hyperbola. A quadratic form may be written $Q(x, y) = \mathbf{XAX}'$, where **X** is a row vector, **X'** its transpose (a column vector), and **A** a symmetric matrix.

The expansion of the expression

$$Q=\begin{bmatrix}x & y\end{bmatrix}\begin{bmatrix}4 & 3\\3 & 8\end{bmatrix}\begin{bmatrix}x\\y\end{bmatrix}=\begin{bmatrix}x & y\end{bmatrix}\begin{bmatrix}4x+3y\\3x+8y\end{bmatrix}=4x^2+3xy+3yx+8y^2$$

shows that the diagonal of **A** contains the coefficients of x^2 and y^2, while the coefficient of xy is the sum of the remaining elements in the symmetric ($a_{ij}=a_{ji}$) matrix.

By rotating the coordinate system through a suitable angle we may eliminate the cross product to obtain the basic form

$$\alpha x_1^2+\beta y_1^2=1 \quad \text{or} \quad x_1^2/a^2+y_1^2/b^2=1$$

in terms of the new coordinates (x_1, y_1). From the signs of α and β it is then a simple matter to decide which type of curve we are dealing with.

The key to this transformation is

```
A=[4 3; 3 8],  [C, E]= eig(A)
```

which gives the answer

```
C =
     0.8817      0.4719
    -0.4719      0.8817

E =
     2.3944      0
     0           9.6056
```

Linear algebra teaches us that the eigenvalues contained in E are nothing else than the coefficients α and β. In our particular case they become 2.3944 and 9.6056, and since they are of the same sign the conic section must be an ellipse.

The relation between the principal coordinates (x_1, y_1) of the ellipse and the original ones may be obtained by the matrix transformation

$$\begin{bmatrix}x\\y\end{bmatrix}=\mathbf{C}\begin{bmatrix}x_1\\y_1\end{bmatrix}$$

We may thus find a principal axis in the (x, y) system by first putting $x_1=0$. From the numeric matrix C we thus obtain

$$\begin{bmatrix} x \\ y \end{bmatrix} = \begin{bmatrix} 0.8817 & 0.4719 \\ -0.4719 & 0.8817 \end{bmatrix} \begin{bmatrix} 0 \\ y_1 \end{bmatrix}$$

which is equivalent to

$$\begin{cases} x = 0.4719 y_1 \\ y = 0.8817 y_1 \end{cases}$$

or

$$y = 1.868x$$

and this is the answer to our problem. Putting $y_1 = 0$ instead gives us the equation for the other axis ($y = -0.5352x$).

Using the above values in $\alpha = 1/a^2$ and $\beta = 1/b^2$ we immediately obtain the lengths a and b of the principal axes. The ellipse and its axes are shown in the next figure.

It is possible to apply a similar procedure to surfaces in (x, y, z) space.

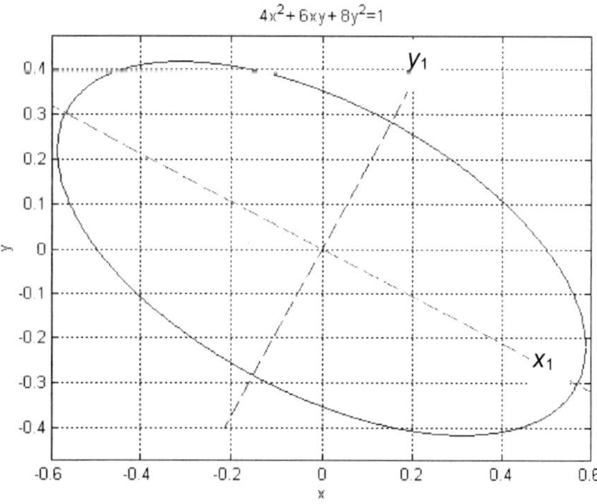

Exercises

❑ Find the value of the determinant and solve the homogeneous system

$$\begin{cases} 2x + 3y = 0 \\ 4x + 6y = 0 \end{cases}$$

by the symbolic program. Use y to normalize the solution vector.

❑ Generate a random, symmetric matrix of size 5×5. For each eigenvalue, calculate the corresponding eigenvector and display that as a row vector. As the next step, calculate the magnitude of each vector. Finally, calculate the determinant of each homogeneous system of equations.

❑ Solve the equation for the above ellipse with respect to y and plot the two branches using *axis equal*. Compare the lengths of the axes and the slopes to that obtained from the eigenvalue calculations.

❑ As an alternative to the eigenvalue method, study the ellipse in polar coordinates (p.59) by the symbolic program. First substitute r*cos(t) for x in the equation, then similarly for y. Next, solve the new equation to obtain the radius r as a function of the polar angle t. Finally, plot r(t) and determine the angles and radii corresponding to the extreme points. Compare to the previous numeric results.

❑ Transform the equation $4x^2 + 6xy - y^2 = 1$ into a coordinate system where the cross term vanishes. What type of conic section is this?

❑ Define a 2×2 matrix A by the symbols a_{11}, etc., and a column vector by [x1 y1]. Find the transformation matrix C corresponding to A by means of *eig*. Execute XY=C*[x1 y1] to find the column vector consisting of x and y. Having obtained the relations between the coordinates, find and admire the (complicated) equations for the principal axes x1 and y1 in terms of the matrix elements.

Conclusion

While working your way through this volume, you have probably found that

- it is easy to demonstrate the principles of undergraduate mathematics using a small subset of MATLAB.
- the examples in this book cover the essentials of practical mathematics for science undergraduates.
- the examples introduce a small set of easy numerical algorithms that are useful in future applications.
- this course gives valuable practical skill in using MATLAB for numerical calculations.
- a student can use the symbolic part of MATLAB to eliminate time-consuming analytical exercises in mathematics.
- the time allotted to conventional mathematics for scientists may be reduced to less than one-half of what is customary at present.

Academic teachers and students who do *not* agree with some of the above points are invited to discuss these matters by email under the address

gunnar.backstrom@physics.umu.se

Index

Vocabulary of MATLAB